Analytical Chemistry

Skill Enhancement Course

Under New Choice Based Credit System (CBCS) for
BSc (Honours and General | Programme) Courses in all Indian Universities

Analytical Chemistry

Skill Enhancement Course

Under New Choice Based Credit System (CBCS) for
BSc (Honours and General|Programme) Courses in all Indian Universities

Krishna Chattopadhyay
MSc, PhD
Postdoctoral Fellow
Department of Chemistry
University of Calcutta
Kolkata

Manas Mandal
MSc, MTech
Assistant Professor
Department of Chemistry
Sree Chaitanya College, Habra
Kolkata
(West Bengal State University)

CBS

CBS Publishers & Distributors Pvt Ltd

New Delhi • Bengaluru • Chennai • Kochi • Kolkata • Lucknow • Mumbai
Hyderabad • Jharkhand • Nagpur • Patna • Pune • Uttarakhand

Analytical Chemistry

ISBN: 978-93-90709-90-8

Copyright © Authors and Publisher

First Edition: 2022

Published by Satish Kumar Jain and produced by Varun Jain for

CBS Publishers & Distributors Pvt Ltd
4819/XI, Prahlad Street, 24 Ansari Road, Daryaganj, New Delhi 110 002, India
Ph: 011-23289259, 23266861, 23266867 Fax: 011-23243014   Website: www.cbspd.com
e-mail: delhi@cbspd.com; cbspubs@airtelmail.in

Corporate Office: 204 FIE, Industrial Area, Patparganj, Delhi 110 092
Ph: 011-4934 4934   Fax: 011-4934 4935   e-mail: publishing@cbspd.com; publicity@cbspd.com

Branches

- **Bengaluru:** Seema House 2975, 17th Cross, K.R. Road,
  Banasankari 2nd Stage, Bengaluru 560 070, Karnataka
  Ph: +91-80-26771678/79   Fax: +91-80-26771680   e-mail: bangalore@cbspd.com
- **Chennai:** 7, Subbaraya Street, Shenoy Nagar, Chennai 600 030, Tamil Nadu
  Ph: +91-44-26680620, 26681266   Fax: +91-44-42032115   e-mail: chennai@cbspd.com
- **Kochi:** 42/1325, 1326, Power House Road, Opp KSEB Power House,
  Ernakulam 682 018, Kochi, Kerala
  Ph: +91-484-4059061-65   Fax: +91-484-4059065   e-mail: kochi@cbspd.com
- **Kolkata:** 147, Hind Ceramics Compound, 1st Floor, Nilgunj Road, Belghoria, Kolkata 700 056, West Bengal, India
  Ph: +91-9096713055, 7798394118, 9836841399   e-mail: kolkata@cbspd.com
- **Lucknow:** Basement, Khushnuma Complex, 7-Meerabai Marg (behind Jawahar Bhawan), Lucknow 226 001, UP
  Ph: +91-522-400043, 9919002738   e-mail: tiwari.lucknow@cbspd.com
- **Mumbai:** PWD Shed, Gala no. 25/26, Ramchandra Bhatt Marg, Next to JJ Hospital Gate no. 2
  Opp. Union Bank of India, Noorbaug, Mumbai-400009, Maharashtra, India
  Ph: 022-60061880/89   e-mail: mumbai@cbspd.com

Representatives

| • Hyderabad | 0-9885175004 | • Jharkhand | 0-9811541605 | • Nagpur | 0-9421945513 |
| • Patna | 0-9334159340 | • Pune | 0-9623451994 | • Uttarakhand | 0-9716462459 |

Printed at: Glorious Printers, Daryaganj, Delhi, India

to

Our parents
and
Aishani, Ishaan, Mainak, Ankan

Preface

"I am among those who think that science has great beauty. A scientist in his laboratory is not only a technician: He is also a child place before natural phenomenon, which impress him like a fairy tale."

Marie Curie (1867–1934)

It gives us immense pleasure in presenting the book *Analytical Chemistry* to all BSc (Honours and General | Programme) students in all the universities in India for their skill enhancement course (SEC). This book is written based on the newly implemented UGC syllabus under choice-based credit system (CBCS) curriculum.

Science is not complete without analysis and one should have the knowledge of analytical chemistry to do the proper analysis. This book includes a detailed discussion on various topics of analytical chemistry, such as analysis methods of soil, water, foods, and cosmetics. Different chromatographic techniques are discussed in detail. The instrumentation of different spectroscopic techniques, such as flame photometer, spectrophotometers is well described with schematic diagram which makes it easy to understand to the readers. The experimental methods in practical part are also well explained in a comprehensible manner.

Following salient features of the book will make it reader-friendly:

- Brief and to-the-point discussion on different sections of analysis increases curiosity among the readers and encourages the students to learn quickly.
- Stepwise instructions for practical experiments help the reader to conduct the experiment smoothly.
- Each chapter includes multiple choice questions (MCQs) and practice questions which will be helpful for the students to prepare for the university as well as different competitive examinations.

We believe that this book will be very useful for the students who even face difficulties to understand this course. Constructive views, suggestions and comments from the readers are most welcome for the improvement of the book.

Krishna Chattopadhyay
Email: krishchem001@gmail.com
Manas Mandal
Email: manasmandal26@gmail.com

Acknowledgements

We are thankful to Prof. Debasish Ray (IIT, Kharagpur), Prof. Swapan Kumar Bhattacharya (Jadavpur University), Prof. Tapan Kanti Paine (IACS, Kolkata) and Prof. Dilip Kumar Maiti (Calcutta University) to their endless encouragements, continuous support and help in acquiring our knowledge we needed to take the next steps toward our dream of writing a book.

We are grateful to all the faculty members of the Department of Chemistry, Sree Chaitanya College, Habra, Kolkata, and all the staff of the college library for their help and inspiration in various ways while writing this book.

We are especially indebted to our parents. Their blessings and love have inspired us to write this book. We express our sincere love to our beloved son Ishaan. Sometimes, we have had to write this book depriving him of his due affection. Hopefully, when he grows up he will understand the importance of our work.

We would like to express our eternal gratitude to our family members for their everlasting love and support. We are also thankful to all of our friends who helped us in every possible way.

We are especially grateful to Mr YN Arjuna (Senior Vice President—Publishing, Editorial and Publicity), Ms Ritu Chawla (AGM—Production) and all the staff of CBS Publishers & Distributors Pvt Ltd for their unwavering support in publishing this book.

The last, but not the least, we thank God for showering His blessings and grace for moving us forward and conceding us the potential to complete the book successfully.

Krishna Chattopadhyay
Manas Mandal

Contents

Syllabus

For BSc (Honours and General | Programme Courses) under CBCS Curriculum of UGC, India [For Delhi University, Calcutta University, West Bengal State University, Burdwan University, Bankura University, etc.]

Skill Enhancement Course
BASIC ANALYTICAL CHEMISTRY
(Credits: 02) 30 Lectures

Introduction: Introduction to analytical chemistry and its interdisciplinary nature. Concept of sampling. Importance of accuracy, precision and sources of error in analytical measurements. Presentation of experimental data and results, from the point of view of significant figures.

Analysis of soil: Composition of soil, concept of pH and pH measurement, complexometric titrations, chelation, chelating agents, use of indicators
  a. Determination of pH of soil samples.
  b. Estimation of calcium and magnesium ions as calcium carbonate by complexometric titration.

Analysis of water: Definition of pure water, sources responsible for contaminating water, water sampling methods, water purification methods.
  a. Determination of pH, acidity and alkalinity of a water sample.
  b. Determination of dissolved oxygen (DO) of a water sample.

Analysis of food products: Nutritional value of foods, idea about food processing and food preservations and adulteration.
  a. Identification of adulterants in some common food items like coffee powder, asafoetida, chilli powder, turmeric powder, coriander powder, pulses, etc.
  b. Analysis of preservatives and colouring matter.

Chromatography: Definition, general introduction on principles of chromatography, paper chromatography, TLC, etc.
  a. Paper chromatographic separation of mixture of metal ion (Fe^{3+} and Al^{3+}).
  b. To compare paint samples by TLC method.

Ion-exchange: Column, ion-exchange chromatography, etc.
  Determination of ion exchange capacity of anion/cation exchange resin (using batch procedure if use of column is not feasible).

Analysis of cosmetics: Major and minor constituents and their function

a. Analysis of deodorants and antiperspirants, Al, Zn, boric acid, chloride, sulphate.

b. Determination of constituents of talcum powder: Magnesium oxide, calcium oxide, zinc oxide and calcium carbonate by complexometric titration.

Suggested Applications (Any one):

a. To study the use of phenolphthalein in trap cases.

b. To analyze arson accelerants.

c. To carry out analysis of gasoline.

Suggested Instrumental demonstrations:

a. Estimation of macronutrients: Potassium, calcium, magnesium in soil samples by flame photometry.

b. Spectrophotometric determination of iron in vitamin/dietary tablets.

c. Spectrophotometric identification and determination of caffeine and benzoic acid in soft drink.

Abbreviations

2D	Two-dimensional
3D	Three-dimensional
AAS	Atomic absorption spectroscopy
Aq.	Aquous
ASAP	As soon as possible
AW	Atomic weight
BOD	Biological oxygen demand
COD	Chemical oxygen demand
Conc.	Concentrate
DI	Deionized
Dil.	Dilute
DO	Dissolved oxygen
EBT	Eriochrome black T
EDTA	Ethylenediaminetetraacetic acid
GC	Gas chromatography
HPLC	High performance liquid chromatography and Applied Chemistry
IUPAC	International Union of Pure and Applied Chemistry
MP	Melting point
MW	Molecular weight
PE	Polyethylene
pH	Potential of hydrogen
ppb	Parts per billion
ppm	Parts per million
RB	Round bottom
RH	Relative humidity
RO	Reverse osmosis

RSD	Relative standard deviation
SRM	Standard reference material
TLC	Thin layer chromatography
tsp	Table spoon
UV	Ultraviolet
VLC	Vacuum liquid chromatography
WHO	World Health Organization

1

Introduction to Analytical Chemistry

UGC Syllabus

Introduction to analytical chemistry and its interdisciplinary nature. Concept of sampling. Importance of accuracy, precision and sources of error in analytical measurements. Presentation of experimental data and results, from the point of view of significant figures.

INTRODUCTION

Analytical chemistry is a branch of chemistry which uses several methods and instruments to identify an unknown chemical substance quantitatively as well as qualitatively. Qualitative analysis gives the idea of chemical identity of the substance, whereas the quantitative analysis measures the relative quantities of the substances in numerical value. Analytical measurements are applied to solve the real problems in wide areas of science such as in biology, geology, medicine, environmental sciences, agriculture, material science, medicine, etc.

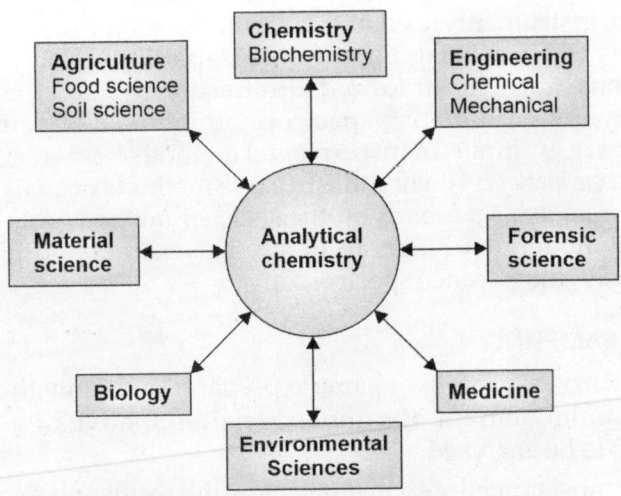

Fig. 1.1: Interdisciplinary nature of analytical chemistry

Interdisciplinary nature: The interdisciplinary nature of analytical chemistry promotes it as an essential tool for all kinds of laboratory (Fig. 1.1). An analytical chemist uses it to solve the real practical problems. Scientific analysis cannot be completed without the help of analytical chemistry.

(i) Analytical chemistry has huge applications in agricultural field. Such as, for obtaining maximum production, a farmer needs to know first the nature of soil (soil pH), the availability of essential nutrients, etc. Thorough analysis of the soil is performed using the knowledge of analytical chemistry. Fertilizers are also analysed before their administration to the soil.

(ii) The food items and many other consumable products (cosmetics and medicines) are analysed by different techniques of analytical chemistry such as gravimetric and volumetric analysis (complexometric titrations, acid–base titrations), various types of chromatographic methods, spectroscopic techniques, etc.

(iii) The medicinal science is also dependent on analytical chemistry. Not only the composition analysis of medicine, but also the examination of different samples like blood, serum, urine, etc. is done using various analytical methods.

(iv) This branch of chemistry is an indispensable part of metallurgy. Analysis of ores, extraction of metal ions and their quality checking is performed by analytical chemistry.

(v) Analytical chemistry is also employed in forensic sciences where different chemicals and instrumental methods are used to prove a criminal act.

1.1 ANALYTICAL METHODS

The most important step of any analysis is choice of proper analytical methods. The analytical method is selected depending on the number and nature of samples, desired accuracy, etc. Analytical methods are mainly classified into three types.

(i) *Classical method:* It is the simplest and cheapest method of analysis where the required chemicals and apparatus are readily available in any chemical laboratory. Calibration of known sample is not required in this method. Titrimetry and gravimetry are two classical methods of analysis.

(ii) *Instrumental method:* This method is more expensive and faster than classical method. The instruments used in this method are very sophisticated, expensive and sensitive. These instruments are capable of determining very low concentrations (trace, substrace and ultratrace analysis) of a chemical substance. Colorimetry, potentiometry, spectrophotometry, voltammetry, chromato-graphy, etc. are examples of instrumental methods.

(iii) *Non-destructive method:* The non-destructive methods of analysis are very costly and speed of analysis depends on the selected method. This method produces results with moderate to high accuracy. The example of this type of method includes X-ray fluorescence spectroscopy.

1.2 CONCEPT OF SAMPLING

In analytical chemistry, sampling or sample preparation is a method of collecting or extracting an optimum amount of sample as representative of a larger amount of substance which is to be analysed.

Sampling is the most crucial step in analysis as the result and its accuracy depends on it. Depending up on the state of a substance, i.e. heterogeneous or homogeneous, solid, liquid or gas, the sampling procedure is different. Following are some important terminologies which readers should know about the sampling procedure in analytical chemistry.

• *Homogeneous substance:* For homogeneous material, a *gross sample* is taken for the analysis such as in clinical laboratory blood, urine, etc. are analysed directly.

- *Heterogeneous substance:* For heterogeneous material, a combination of materials is collected to make a *gross sample*.
- *Gross sample:* It contains several portion of the substance to be analysed. The amount of the sample is within the range of few grams to kilograms.
- *Laboratory sample:* It contains a small portion of homogeneous **gross sample**. The amount of the sample is taken as few grams.
- *Analysis sample:* It is the actual amount of sample that is analysed. The amount of the sample is taken as few milligrams or millilitre.

The sampling techniques involving solid, liquid and gas samples are discussed briefly.

Solid sampling: Sampling of the solid sample is most difficult due to its less homogeneous nature. For this reason, solids are crushed or ground into smaller particles to a homogenized sample. Depending on the hardness of the material automated or manual crusher or grinder is used. Polymer pellets are treated with liquid nitrogen, which freezes the polymer to make it brittle in nature and then these polymers are easily grinded to obtain the powder form. Soft materials cannot be easily grinded as they just deform rather than crushing. Therefore, soft materials are often dried before grinding using oven to remove any adsorbed liquids in order to get the representative sample. A large number of analytical techniques employing grinding, blending and pulverizing instruments are used for sampling the solid substances such as soil, cement, ceramics, food, textile and other solid materials.

Liquid sampling: For sampling a liquid, portions of the same are collected from the sources and treated according to the nature of analysis. In general, liquid samples are first stirred or shaken well before considering it as a *grab sample*. For the preparation of a *representative sample* from big sources (such as analysis of lake water or river water), samples are collected from different locations or layers. Then these samples are either analysed separately or they can be blended together to obtain a representative sample. To analyse the normal river water, the samples are collected away from contaminates such as floating froth, river banks, municipal waste treatment sites and industrial waste disposal areas. Sometimes, centrifugation or filtration is performed to remove solid particles present in liquid sample. For liquid samples having two immiscible layers, either separate layer analysis or emulsification method is adopted for sampling. Liquid suspensions like milk, orange juice, antacids are sampled as it is without the removal of the suspended solids.

Gas sampling: Gas samples are homogeneous in nature but they may often form separate layers of different density. Therefore, gas samples are needed to be stirred prior to analysis. Gas samples are collected either at a particular point at a time or a batch of samples are collected at different locations over a certain time period and then mixed together to obtain the average representative sample. For collection of gas samples, containers like balloons, gas tight syringes, plastic bags, etc. can be used. Suitable PPE (personal protective equipment) should be used for sampling of toxic or flammable gases. Gas samples can also be collected by absorbing the gas into a solid or liquid adsorbent. This process is called 'scrubbing of gas'. Scrubbing also reduces the sample size which makes it easy to carry. For example, organic pollutant vapour present in air can be collected in an activated charcoal bead of size as small as a ball point pen. The gas samples can often be filtered to remove any undesirable solid particles present in the sample.

1.3 ERRORS IN CHEMICAL ANALYSIS

Almost every measurement in chemical analysis involves some uncertainties, which is defined as errors. The error can be classified into two principal types such as *determinate* or *systematic error* and *indeterminate* or *random error*.

(I) **Determinate or systematic error:** This type of errors is mainly caused by the design of an improper analytical procedure or faults in the equipment used in the analysis. The name determinate suggests that this type of error may be determined and hence it is either avoided or corrected through the proper analysis. Determinate errors influence the accuracy of the measurement.

Depending upon the *relative magnitude of the error with respect to the size of the sample*, the determinate errors are of two types:

- **Constant errors:** The magnitude of constant error does not change with the change in sample size. Thus, the effect of this type of error is more prominent when the sample size is decreased. Let us consider three samples 20, 35 and 50 mg/L of analyte for which the measured values are 25, 40 and 55 mg/L, respectively. This analysis has a constant positive error of 5 mg/L with the relative percentages of constant errors 25%, 14% and 10%, respectively, for the three measured values. Therefore, the effect of constant error can be reduced by increasing the sample size.
- **Proportional errors:** The magnitude of proportional error varies depending upon the size of the sample. With this type of errors, the absolute error changes with the size of the sample, but the relative error remains constant. Let us consider the same three samples 20, 35 and 50 mg/L of analyte where the measured values are 21.6, 37.8 and 54 mg/L, respectively. This analysis has a proportional positive error of 1.6, 2.8, 4 mg/L, respectively. The method results in a constant relative error of 8% for the three measured values.

Based on the *sources*, the determinate or systematic errors are mainly of three types:

- **Instrumental errors:** The effect of improper calibration or the environmental factors on the instrument or equipment causes the instrumental error. Such as measuring equipment like burette, pipet, measuring cylinder, volumetric flask, etc. does not provide the correct volume in all environmental condition. An electronic instrument gives this error due to fluctuation of current voltage, incorrect calibration, changes in temperature, etc.
- **Methodical errors:** The imperfect chemical or physical behaviour of the reagents and reactions such as very slow reaction rate, low specificity of reagents, contaminated or low-grade reagents, instability of some products, unwanted side reactions, etc. affect the analysis and produces the systematic method errors. For example, in titrimetric method an excess (though very small) volume of titrant is required for the colour change of indicator used.
- **Personal or analyst error:** This type of errors is introduced in the analysis results due to carelessness, ignorance or inattention of the analyst or the person who is performing the analysis. Knowingly or unknowingly, use of an incorrect setting in an instrument or the wrong operating condition may lead to the analyst error.

However, with proper setting, an analyst also can do some common errors like transcription errors (by noting down the wrong data on lab notebook) and calculation error.

Minimizing the systematic errors in analytical methods: The complete elimination of the systematic error is very challenging task. However, this type of error can be identified and minimized with proper choice of the analytical methods and operating conditions. The following measures can be taken to identify and reduce systematic errors in an analytical method:

a. *Calibration of instruments:* Systematic error can often be detected by comparison to theoretical model or someone else's result. However, in such case, it is difficult to tell which one is accurate. Calibration (if feasible) is the most trustworthy approach to decrease systematic errors. A reference quantity, for which the actual result is already known, can be used to calibrate the analytical procedure. However, if possible, it is always better to calibrate the whole procedure and apparatus, on a known quantity of sample of similar size to your unknown quantities.

b. *Analysis of standard samples:* The best way to measure the error originating in an analytical method is to analyse standard reference materials (SRMs). The SRMs contain one or more analytes in known concentration. The standard reference materials can either be commercially purchased or synthetically produced in the laboratory. For preparation of SRMs, utmost care should be taken so that the overall composition of the synthetic standard material closely approaches the composition of the samples to be analysed.

If results of analysis on the SRMs differ from the accepted value, one should check with the random error (if any) associated with the analytical method.

c. *Variation in sample size:* The extent of determinate error is dependent on the sample size. With increase in the size of sample, the effect of constant error decreases. Hence, multiple measurements with varying sample size can often be helpful in determination of constant error.

d. *Blank determinations:* Blank comprises of all the reagents except the analyte in the same solvent used in a determination. Often some reagents are added to the sample to create the analyte environment, which is known as sample matrix. In a blank measurement, all steps of the analysis are executed on the matrix without the analyte. The results of balnk determination are then applied for correction of the analyte measurement in order to remove errors originating from unwanted contaminants present in the reagents and vessels used for analysis.

e. *Independent analysis:* To compare the results and to avoid some common sources of errors, a second independent analytical method which differs significantly from the existing method can be performed in parallel. Generally, it is very much useful in case of non-availability of SRMs.

(II) *Indeterminate or random error:* This type of errors originates from the uncontrollable variables in the analysis. It is generally small but irregular and cannot be corrected. A better measurement can help to reduce it yet. Indeterminate errors influence the precision of the measurement.

Minimizing random errors in analytical methods: The occurrence of random error in a measurement is very irregular and has no preferred direction. Therefore, it cannot be eliminated from the measurement, however, following measures can be taken in order to minimize the effect of random error:

(a) Large number of measurements can be performed to obtain an average value. This method reduces the effect of random error.
(b) Imprecision, i.e. the impact of random error can be reduced by large sample sizes.
(c) Maintaining appropriate experimental method can also be helpful in reducing the effect of random error.

1.4 ACCURACY AND PRECISION

- *Accuracy:* Accuracy is defined as the nearness of a measured value to the actual value. It signifies the parity between a result and the accepted value. The smaller the difference between these two values, the greater is the accuracy (Fig. 1.2). The greater is the value of accuracy, the more correct is the analysis or measurement. In most of the cases actual true value is not known, it is difficult to determine the accuracy and accepted true value is used in place of true value.

  Accuracy can be mathematically expressed by two terms—*absolute error* and *relative error*.

  – *Absolute error (E_a):* It is the difference between the measured or experimental value and the true value of a quantity and it is expressed as: $E_a = x_i - x_t$, where x_i is the measured and x_t is the true value. The E_a has the unit dependent on the type of measurement.

    Example: Let, the 28.31 cm is the true value of a length measurement and 28.46 cm is the experimental value. Therefore, absolute error = (28.46 – 28.31) cm = 0.15 cm.

  – *Relative error (E_r):* It is described as the ratio of the absolute error (E_a) of the measurement to the true value and it is expressed as $E_r = \dfrac{x_i - x_t}{x_t}$.

    Example: The relative error of the above example will be 0.0053.

    The E_r is dimensionless and it is more significant or better measurement than absolute error. The percent relative error can be expressed by multiplying the fractional error by 100.

- *Precision:* It is referred to the extent of closeness between the results of two or more replicate measurements on the same system using identical method of analysis. It indicates the reproducibility of the method (Fig. 1.2).

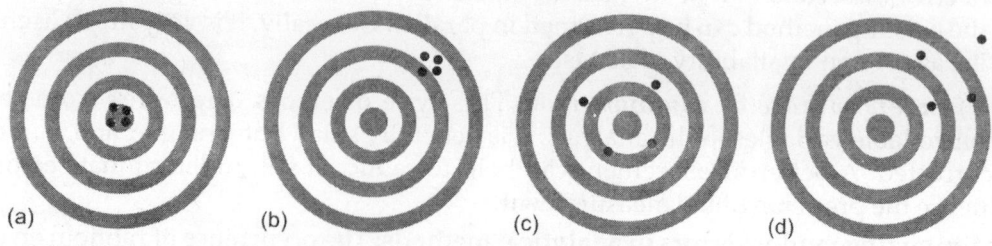

Fig. 1.2: (a) High accuracy and precision, (b) low accuracy but high precision, (c) high accuracy but low precision, (d) low accuracy and precision

For example, if replicate measurements of a data by method A gives the values 5.35, 5.32 and 5.36 while method B gives the values 6.32, 6.38 and 6.42, precision of method A is better than that of method B.

The precision of a result obtained in a set of measurement is expressed in terms of the *average deviation of the mean, relative average deviation (RAD), standard deviation (SD)* and *relative standard deviation (RSD)*.

– *Average deviation of the mean:* Let N numbers of measurements are performed for quantity 'a'. The measured values are $x_1, x_2, x_3, ..., x_N$ and the mean value is $\bar{x}$. Then the deviations of the individual values from the mean values are $(x_1 - \bar{x}), (x_2 - \bar{x}), (x_3 - \bar{x}), ..., (x_N - \bar{x})$.

Then the mean deviation

$$= \frac{|x_1 - \bar{x}| + |x_2 - \bar{x}| + |x_3 - \bar{x}| + ...|x_N - \bar{x}|}{\bar{x}} = \frac{\sum_{i=1}^{N} |x_i - \bar{x}|}{N}$$

The average deviation of the mean is obtained by dividing the mean deviation by square root of the number of measurements.

Therefore, the average deviation of the mean

$$= \frac{\text{Mean deviation}}{\sqrt{N}} = \frac{\sum_{i=1}^{N} |x_i - \bar{x}|/N}{\sqrt{N}}$$

– *Relative average deviation:* It is expressed in either of the three following terms:

  I. Relative average deviation (%)

$$= \frac{\sum_{i=1}^{N} |x_i - \bar{x}|}{x \cdot N} \times 100$$

  II. Relative average deviation (parts per thousand or ppt)

$$= \frac{\sum_{i=1}^{N} |x_i - \bar{x}|}{x \cdot N} \times 1000$$

  III. Relative average deviation (parts per million or ppm)

$$= \frac{\sum_{i=1}^{N} |x_i - \bar{x}|}{x \cdot N} \times 10^6$$

– *Standard deviation:* The precision of a measurement is generally expressed in terms of standard deviation. It is defined as the square root of the mean of the squares of individual deviation from their mean value. Generally, it is expressed by the following equation:

Standard deviation

$$= \sqrt{\frac{(x_1 - \bar{x})^2 + (x_2 - \bar{x})^2 + (x_3 - \bar{x})^2 + ... + (x_N - \bar{x})^2}{N}}$$

  I. *Population standard deviation* (σ): It measures the precision of the population (when $N \to \infty$), and expressed as

$$\sigma = \sqrt{\frac{\sum_{i=1}^{N} (x_i - \mu)^2}{N}}$$

where $\mu = \dfrac{\sum_{i=1}^{N} x_i}{N}$ is called the *population mean* and N is the total number of measurements in the population. The μ can never be measured but as N increases, μ approaches to $\bar{x}$.

II. *Sample standard deviation (s):* With small number of sample data (N is small) the extent of precision is measured by the value of sample standard deviation. It is expressed as:

$$s = \sqrt{\frac{\sum_{i=1}^{N}(x_i - \bar{x})^2}{N-1}}$$

where x is the sample mean and $(N-1)$ is called the number of degrees of freedom.

– *Relative standard deviation (RSD):* The relative standard deviation of a measurement is calculated by dividing the standard deviation value with the mean value of all the measurements.

Relative standard deviation (RSD)

$$= \frac{s}{\bar{x}}$$

Percent relative standard deviation (% RSD) or coefficient of variation

$$= \frac{s}{\bar{x}} \times 100\%$$

Relative standard deviation input

$$= \frac{s}{\bar{x}} \times 1000 \text{ ppt}$$

• *%Variance:* The square of the standard deviation value is known as variance (σ^2 or s^2).

Importance of accuracy and precision: To achieve the best quality of measurement, accuracy and precision both are equally important. For a set of measurement, where the true value is not so important, the precise value is more meaningful as they are grouped throughout the series of measurement. On the other hand, where the true value is more important, the accuracy is valued over precision, as it is more useful in measuring the needed value. However, to maintain a measurement system, it should be checked for precision and accuracy regularly, as they are equally important.

1.5 PRESENTATION OF EXPERIMENTAL RESULTS: CONCEPT OF SIGNIFICANT FIGURES

Significant figures: The concept of significant figures is very much important in writing a numerical value of an analytical experimental result. It is defined by all the certain digits in addition with the first uncertain digit in a number. For example, if an analytical balance capable of measuring nearest milligram accurately shows the mass of a sample to be 42.04672 g, then only the first four digits after decimal is meaningful. The last digit known with certainty is 6. The fourth digit 7 is uncertain and it only indicates that the mass is greater than 42.045 g but less than 42.047 g. The fifth or the last digit 2 is meaningless. In another example, while recording a burette reading of a 50 mL burette as shown in Fig. 1.3, one can easily tell that the liquid level is between

30.2 to 30.3 mL. The analyst can also assess the liquid mark between the graduations to about 0.02 mL and report the volume as 30.24 mL. The reported value 30.24 has four significant numbers where first three digits (3, 0, 2) are certain and fourth digit 4 is uncertain.

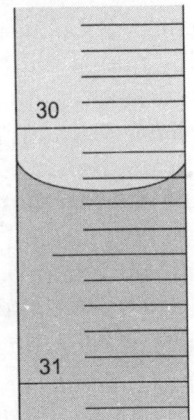

Rules for significant figures:

(i) *All non-zero digits are significant.* For example, the number 35 has two significant figures, 45.2 has three significant figures, 30.45 has four significant figures, 2.5403 has five significant figures.

(ii) A zero becomes significant if it appears in between two non-zero digits. For example, the number 5.034 has four significant figures and 407 has three significant figures.

Fig. 1.3: Graduated burette reading

(iii) When zero appears at first or before non-zero digits, it becomes insignificant. The numbers 0.35 and 0.0035 both have two significant figures.

(iv) Zeros following non-zero digits or placed right to the decimal are significant. For example, 25.0 has three, 25.00 has four, 25.000 has five significant numbers, respectively.

(v) Zero at the end of a whole number becomes significant when it is followed by a decimal. For example, writing 630 means the trailing zero is not significant whereas '630'. Indicates the trailing zero is significant.

(vi) For a number ($M \times 10^n$), all digits included in M are significant according to the above five rules but '10' and n are insignificant. For example, 2.35×10^6 has three significant figures (2, 3, 5).

Rounding off a number: In different analytical methods, the results are generally rounded off to express it to a desired number of significant figures. Before rounding off a number let us know about the different significant figures in detail. The left most digit (do not count a leading zero) is called the most significant digit. If a number is to be rounded off to n significant digits then the nth digit from the most significant digit is called least significant digit. The least significant could be zero. The ($n + 1$)th digit is the first non-significant digit.

Following rules are to be taken care of while rounding off a number.

(i) While rounding off a number, all non-significant digits are dropped.

(ii) When the first non-significant digit is less than 5, the least significant digit remains unchanged.

(iii) When the first non-significant digit is greater than 5, the least significant digit is increased by 1.

(iv) If the first non-significant digit is 5, the least significant number is rounded off to an even digit. While rounding off, zero is considered to be even number.

Example 1: Let us round off 54.2364 to four significant figures. The fifth digit (6) is the first non-significant number which is greater than 5. Therefore, for rounding off, we will drop every figure after the fourth digit (3) and increase it by 1. So, the original number rounds off to 54.24.

Example 2: Let us round off 2.7626 to three significant figures. The fourth digit (2) is the first non-significant figure which is less than 5. Hence we will simply remove all the figures after third digit and the original number rounds off to 2.76.

Example 3: Rounding off 26.235 to four significant figures. The fifth or the first non-significant digit is itself 5. Thus, the fourth digit will be rounded off to an even number and the original number rounds up to 26.24.

Example 4: Rounding off 26.8513 to three significant figures. The fourth or first non-significant digit is 5. The third or last significant digit is 8 which is an even number. So, it will be 26.8 after rounding off.

Rules of rounding off in mathematics:

(i) *For addition or subtraction,* the result is expressed up to the same decimal places as it is in the number with least number of decimal places.

For example, the addition of numbers 125.7, 5.8, 0.637 gives a value of 132.137. Here the number with least decimal place is 5.8. Hence the result should be expressed up to one decimal place. Therefore, the correct answer is 132.1.

Similarly, subtracting 24.36 from 475.264 we get 450.904. So, by the above rule the correct answer will be 450.90.

(ii) *For multiplication or division,* the result is expressed with same number of significant figures as it is in the number with least number of significant figures.

For example, if we multiply 3.425 (four significant figures) with 6.23 (three significant figures), we get 21.33775. By considering the above rule, the result must be reported to three significant figures. Hence, 21.3 is the correct answer.

Similarly, when we divide 8.24534 (six significant figures) by 3.52 (three significant figures), we get 2.34242. By the above rule, the correct result is 2.34.

BIBLIOGRAPHY

1. Harris DC. *Quantitative Chemical Analysis* (7th Edn.), 2007, by WH Freeman and Company.
2. Robinson JW, Frame ES and Frame II GM. *Undergraduate Instrumental Analysis* (6th Edn.), 2005, by Marcel Dekker.
3. Skoog DA, Donald MWF, James H Stanley. *Fundamentals of Analytical Chemistry* (9th Edn.), 2013, Cengage Learning.

QUESTIONS

Multiple Choice Questions

1. Analytical chemistry is employed for:
   - (a) Quantitative analysis
   - (b) Qualitative analysis
   - (c) Both a and b
   - (d) None of these
2. Qualitative analysis gives information about:
   - (a) Amount of the sample
   - (b) Nature of the sample
   - (c) Both a and b
   - (d) None of these
3. Instrumental error is which type of the following errors?
   - (a) Determinate
   - (b) Indeterminate
   - (c) Random
   - (d) None of these
4. The proper way to minimize the random error is:
   - (a) Changing the instrument
   - (b) Taking repeated measurements to get an average value

(c) Taking help from other analyst

(d) None of these

5. The nearness of a measured value to the true value is called:

   (a) Mean   (b) Median

   (c) Accuracy   (d) Precision

6. Standard deviation is a measure of:

   (a) Error   (b) Accuracy

   (c) Precision   (d) None of these

7. The number of significant figures in 0.0065 and 65.00 are ...... and ......, respectively.

   (a) 2, 4   (b) 4, 2

   (c) 5, 2   (d) 5, 4

8. After rounding off to three significant figures the number 26.8513 will be:

   (a) 26.9   (b) 26.8

   (c) 26.7   (d) 26.85

Answers

1. (c); 2. (b); 3. (a); 4. (b); 5. (c); 6. (c); 7. (a); 8. (b)

Practice Questions

1. What is analytical chemistry? Discuss its interdisciplinary nature.
2. Discuss different types of analytical methods.
3. What do you mean by sampling? Briefly discuss the sampling for solid and liquid substances.
4. Based on the sources classify the systematic errors and briefly discuss about them.
5. What is the difference between precision and accuracy? Which is more important?
6. What is meant by standard deviation? Derive the mathematical formula for calculating standard deviation.
7. Briefly discuss the concept of significant figures.
8. State and explain the rules for rounding of numerical expression with example.
9. Write down the rules for rounding of while performing mathematical programme.
10. Calculate the standard deviation for the percentage of copper in a sample: 6.23, 6.25, 6.26, 6.24, 6.28 and 6.30.
11. Round of the following numbers to four significant figures: 28.6824, 45.5832, 62.590 and 39.56.
12. What are the absolute and relative errors of the approximation 3.14 to the value of π?

   [Hint: $E_a = |3.14 - \pi| \approx 0.0016$ and $E_x = |3.14 - \pi| / |\pi| \approx 0.00051$]

Chapter

2

Analysis of Soil

UGC Syllabus

Composition of soil, concept of pH and pH measurement, complexometric titrations, chelation, chelating agents, use of indicators.
1. Determination of pH of soil samples.
2. Estimation of calcium and magnesium ions as calcium carbonate by complexometric titration.

INTRODUCTION

Soil, the outermost layer of earth's crust, is the most essential natural resource for the existence of life on this planet. It serves as a medium for the growth of plants. According to Soil Science Society of America (SSSA) Glossary of Soil Science Terms, soil is defined as *"The layer(s) of generally loose mineral and/or organic material that are affected by physical, chemical, and/or biological processes at or near the planetary surface and usually hold liquids, gases, and biota and support plants."* Soil is a living, breathing medium providing support to almost all living beings in the world. The environmental and genetic factors such as climate, macro- and microorganisms acting on the parent material over the years influence the unconsolidated mineral or organic material of soil. Soil and its functions on an ecosystem vary widely from one location to another. As a result, soils of different geographical area differ widely in colour and texture. Thus some soils are black, some are red; some are shallow and some are deep; some are fine-textured and some are coarse-textured.

2.1 COMPOSITION OF SOIL

Soil plays an integral role in providing nutrients to plants and microorganisms living in it. Hence, it is very important to have knowledge of its composition for proper management of the nutrients. The soil has three basic phases (Flowchart 2.1) such as solid phase (minerals or inorganic compounds and organic matters), liquid phase (soil water) and gas phase (soil air). A typical soil comprises of approximately 45% mineral, 5% organic matter, 25% water, and 25% air.

(a) *Solid phase:* This phase mainly consists of inorganic particles and organic matters. The inorganic particles include different minerals along with weathered rock fragments whereas the organic matters (also called, humus) include both living and dead micro- and macroorganisms. The mineral ingredients of solid phase are sand, silt and clay (Fig. 2.1).

Sand: Large particles of diameter 0.02–2 mm. They are rough in texture and individual particles are easily visible. The most common mineral component of sand is silica (SiO_2). Particles larger in size than sand are not considered as soil.

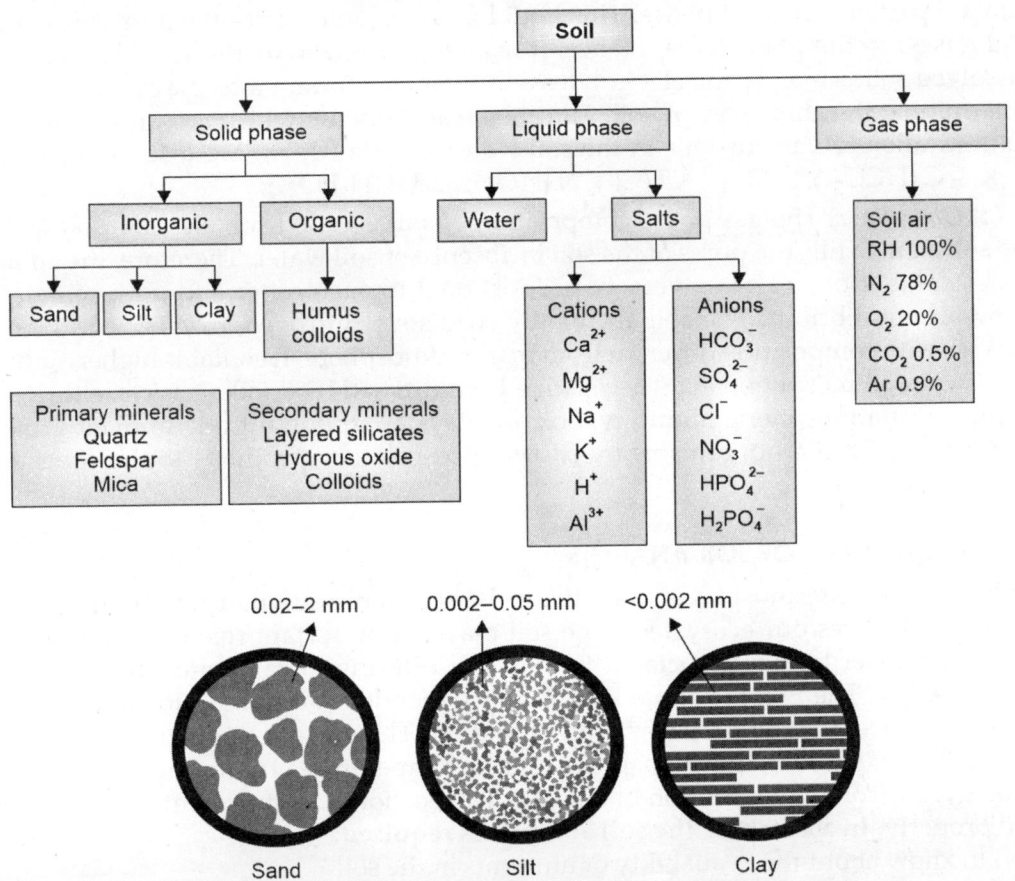

Flowchart 2.1: Composition of soil

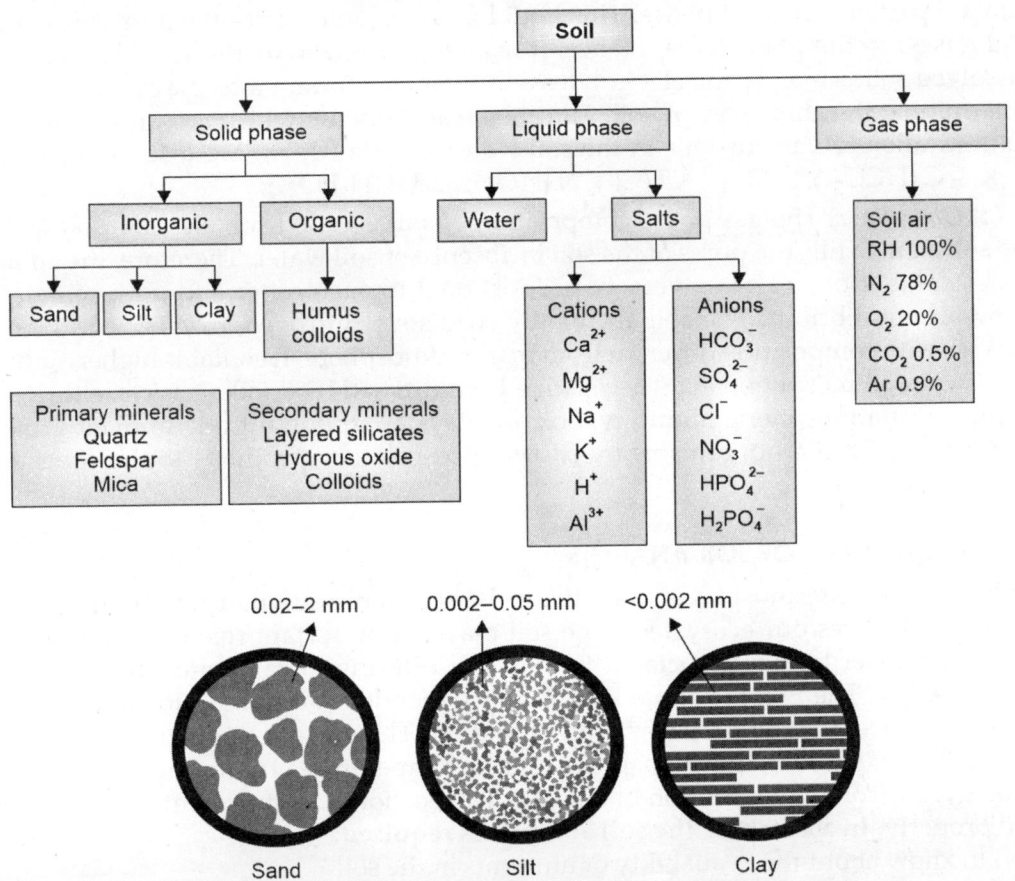

Fig. 2.1: Solid phase (sand, silt and clay) of soil

Silt: Silt is constituted by medium-sized particles of diameter 0.002–0.05 mm.

Clay: Small particles of diameter <0.002 mm are known as soil colloids or clay.

Types of soil: Depending upon the relative composition of solid phases (sand, silt, clay), soils are classified into three basic types—sandy soil, clayey soil and loamy soil.

(i) *Sandy soil:* This type of soil mainly consists of sand and no other particles such as clay or silt. Due to bigger size of sand particle, it has large pores within it. Therefore, sandy soil cannot hold enough water but it provides good aeration. This type of soil is not suitable for plant growth. However, an addition of humus can make it fertile.

(ii) *Clayey soil:* The name defines that this type of soil mainly consists of clay. Due to close packing of small clay particles, it has very small pores present within it. Therefore, clayey soil cannot trap enough air but it tightly holds water. However, this type of soil is rich in mineral content which is very essential for plant growth. Although clayey soil is more fertile than sandy soil, an addition of humus and sand makes it superior for plant growth.

(iii) *Loamy soil:* This type of soil consists of sand, clay, silt and humus in optimum ratio. Therefore, it can hold enough water; has adequate aeration, as well as high mineral content, providing the necessary nutrients for the growth of plants.

(b) *Liquid phase:* Water often called soil water fills the available pores between the mineral particles of solid phase. This liquid phase of soil carries the nutrients (solids and gases) to the plant roots. Hence, it is also referred to as the soil solution. The dissolved solids mainly the electrolytes and their constituent ions play crucial role in determining the chemistry of soil. The chemical elements which are present in high concentration as various ions in the soil water are: Ca (Ca^{2+}), Mg (Mg^{2+}), Na (Na^+), K (K^+), C (HCO_3^-), S (SO_4^{2-}), Cl (Cl^-), N (NO_3^-), and P ($H_2PO_4^{2-}$).

(c) *Gas phase:* The gas phase composition of soil is called soil air. Like soil water, the soil air also fills the pores of the soil in absence of soil water. Therefore, the air and water content of soil are inversely proportional to each other and both remain in dynamic equilibrium. The soil air mainly consists of nitrogen, oxygen, and carbon dioxide with composition different from that of atmosphere. It contains higher amount of CO_2 than atmospheric air. The relative humidity (RH) of soil air is close to 100%, unlike most atmospheric humidity. Soil air plays an important role in plant growth and activity of microorganisms by carrying respiratory products to the roots and organisms.

2.2 SIGNIFICANCE OF SOIL ANALYSIS

Soil being the outermost layer of earth's crust is a significant natural resource which greatly influences our ecosystem. The soil plays an important role in regulating the nature of water bodies associated with it and filtering of rainwater for storage as groundwater. The whole animal kingdom is dependent on the soil for fulfilment of three basic needs—foods, clothing, and building. The soil also regulates the amount and nature of gases in the atmosphere and its temperature. To interpret the climate, geology, hydrology, vegetation of a particular location it is important to analyse the soil properly. **In a nutshell, the soil analysis is required:**

(i) to know about the availability of nutrients in the soil
(ii) to predict the required nutritional values for agriculture
(iii) to enhance the efficiency of fertilisers and water resource inputs
(iv) to reduce the environmental impacts due to soil amendments
(v) to evaluate the fertility status of soil of a particular area.

2.3 INTERPRETATION OF SOIL pH

The soil can be either acidic or alkaline in nature (Fig. 2.2) and the tool used to measure the acidity or alkalinity of a solution is called pH. As the plant growth on a particular

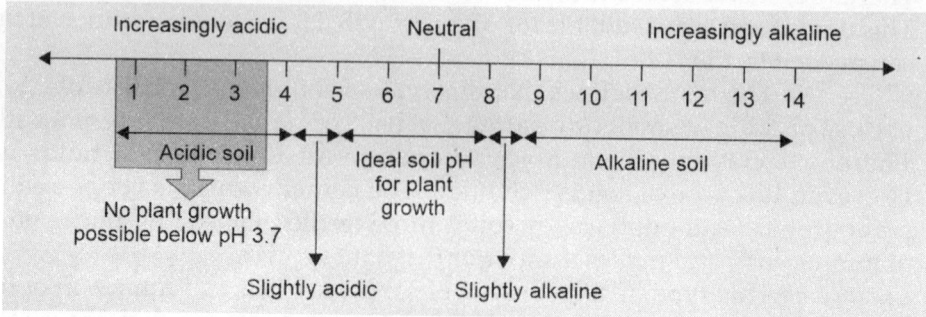

Fig. 2.2: pH dependence of soil

soil is largely dependent on the nature of the soil, it is important to evaluate the soil pH to gain maximum production.

In 1909, Sørensen introduced the term pH which is defined as the negative logarithm of hydrogen ion concentration of a particular solution. The pH scale ranges from 0 to 14 units with the pH = 7 point referred to as neutral. Solutions with a pH >7 are basic in nature whereas solutions with a pH <7 are acidic.

$$pH = -\log[H^+]$$

The soil pH is measured as the negative logarithm of the active hydrogen ion (H^+) concentration in the soil solution. The acidity or alkalinity and thereby the pH of soil is largely controlled by the minerals present in it.

The soil pH has great influence on the availability of different nutrients to the crops growth in that particular soil. It also affects the microbial population in soils. At pH 5.5–6.5 most of the soil nutrients become available to plants and crops. The soil pH also has great influence on the acidity or alkalinity of the water body in contact with it. The water found near the soil rich in minerals like calcite or limestone is basic in nature. On contrary the soil on which mining work is going on or soils rich in minerals like sulphides causes water to be acidic in nature.

Importance of soil pH: The pH value of a soil indicates the suitability of the soil for growth of a particular plant. Thus, for assessment of a soil's nature in order to examine the nutritional availability, physical condition, structure and permeability, the measurement of the pH is an indispensable part.

- The value of soil pH not only provides information about the concentration of the essential nutrients but also tells us the potency of toxic elements present in the soil.
- Soil pH values are also important to assess the status of microorganisms and its effect on the degradation of organic substances and availability of essential nutrients.
- Determination of soil pH is absolutely necessary for soil management for agricultural crops.

Effect of the pH on availability of nutrients in soil: The pH of soil affects the concentration of the nutrients and their relative reactivity (Fig. 2.3). At a low pH, the concentrations of aluminium, iron and manganese in soil increase; sometimes, aluminium and manganese may reach to their toxic levels for plants. However, the useful elements such as calcium, magnesium, phosphorus, and molybdenum become less available. The variations in the nutrients concentrations greatly influence the plant growth. Therefore, knowledge of the soil pH and related acid toxicity is crucial for sowing crops or pastures. In contrast, when the soil pH is greater than 7.5, calcium ions present in soil combine with phosphorus and make it less available to plants. Moreover, increased alkalinity of soil causes zinc and cobalt deficiencies which results in retarded growth of plants and reduced yields of crops and pastures.

Effect of the pH on plant response in soil: The growth of plants is largely affected by the pH of the soil (Table 2.1). Although some plants are tolerant to a wide range of pH, while others are very sensitive to small change of pH. Most of the agricultural plants prefer the soil pH ($CaCl_2$) range of 5.2 to 8.0. The soil pH also influences the microbial activity in the soil. Most of the microbial reactions favour the pH range of 5.0 to 7.0. The extreme soil *pH* (highly acidic or highly alkaline) causes the disappearance of diverse species of earthworms and nitrifying bacteria. The colonization of plant roots by nitrogen-fixing bacteria (*Rhizobia*) is also affected by soil pH and has a preferred pH range of 6.0–7.5 for optimum growth at temperature range of 25–30°C.

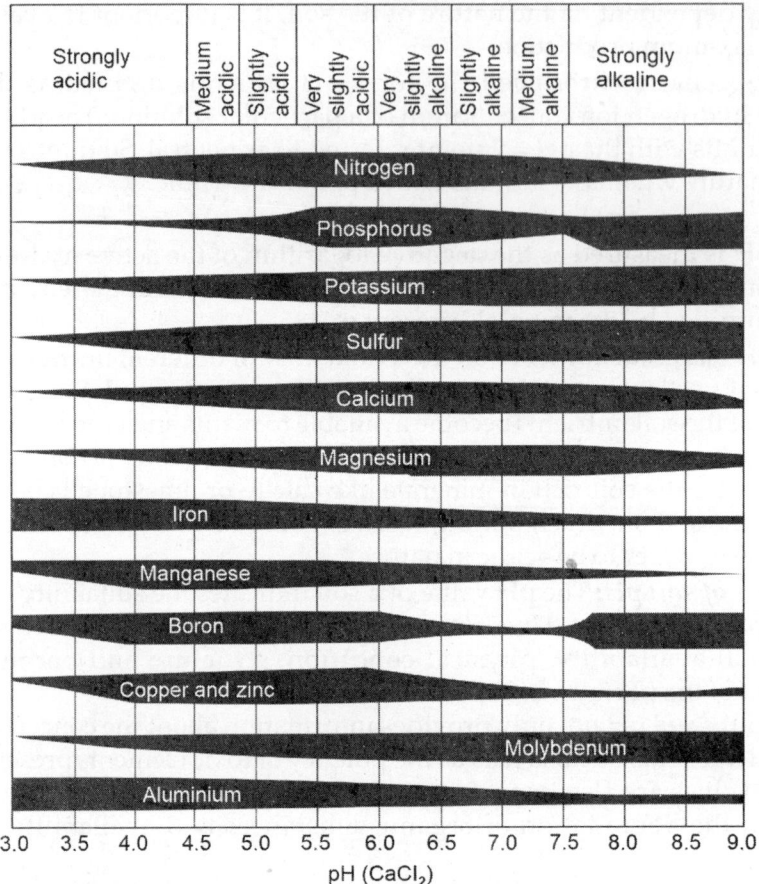

Fig. 2.3: H dependence of available nutrients

The colour of a flower is sometimes defined by the pH of the soil on which they are planted. Big leaf hydrangeas (*H. macrophylla*) are the best example for this. They produce blue colour flowers in acidic soil (pH 5.5 and less than 5.5), whereas in alkaline soil (pH 7 and greater than 7) they give pink flowers. Although at very high pH value of soil (highly alkaline) colour of the flowers becomes white such as 'Lanarth White'.

2.4 DETERMINATION OF pH OF SOIL

Experiment 2.1

Aim: Measurement of soil pH using a potentiometer.

Principle: Using potentiometer, the degree of acidity or alkalinity of a soil sample suspended in DI water and in 0.01 M calcium chloride ($CaCl_2$) solution is determined.

Apparatus: Analytical balance, pH meter, thermometer, magnetic stirrer, stir bar, centrifuge and tubes, pipette, beakers (50 mL, 100 mL), glass rod, sieve, ¼ inch mesh (6.35 mm), mortar and pestle.

Chemical reagents:

   (i) Buffer solutions of pH 4.0, 7.0 and 10.0 (suitably prepared or commercially available).

Table 2.1: The ideal pH ranges of some common fruits and vegetables

pH range	5.0	5.5	6.0	6.5	7.0	7.5	8.0	8.5
Common Fruits								
Pineapple	▓	▓	▓					
Coconut	▓	▓	▓				▓	
Apple		▓	▓	▓				
Mango		▓	▓	▓	▓			
Orange	▓		▓	▓	▓			
Papaya	▓		▓	▓	▓			
Watermelon			▓	▓	▓			
Jackfruit			▓	▓	▓			
Sapodilla		▓	▓	▓	▓			
Banana			▓	▓	▓	▓		
Jamun			▓	▓	▓	▓		
Guava			▓	▓	▓	▓	▓	▓
Common Vegetables								
Potato	▓	▓	▓					
Beans		▓	▓	▓	▓			
Cauliflower		▓	▓	▓	▓			
Cucumber		▓	▓	▓	▓			
Brinjal		▓	▓	▓	▓			
Cabbage			▓	▓	▓			
Pumpkin			▓	▓	▓			
Tomato			▓	▓	▓			
Onion			▓	▓	▓			
Lettuce			▓	▓	▓			
Carrot			▓	▓	▓			
Bottle gourd			▓	▓	▓	▓		
Bitter gourd			▓	▓	▓	▓		
Okra				▓	▓	▓		
Chili			▓	▓	▓			
Peas			▓	▓	▓	▓		
Spinach			▓	▓	▓	▓	▓	

(ii) *0.01 M CaCl₂ solution:* Dissolve 1.47 g $CaCl_2 \cdot 2H_2O$ in 1 L DI water. The pH of this solution should be 5.0–6.5. If required adjust pH with $Ca(OH)_2$ or HCl.

Sample preparation:

(i) *With water:* Collect the soil sample and air dry it. Break the bigger soil lumps and grind it with a mortar-pestle. Separate the finer soil sample by sieving. Weigh 10 g of the fine soil particles into a 50 mL beaker and add 10 mL of DI water.Using a glass rod, mix it thoroughly. Then stir the mixture using a magnetic stirrer until the soil is fully suspended in water. Let the mixture stand for 10 min to settle.

Repeat the stirring and settling process for three times. Measure the pH of the supernatant using pH meter.

(ii) *With calcium chloride solution:* Follow the above procedure to prepare another soil extract using 10 mL 0.01 M $CaCl_2$ aqueous solution in place of DI water.

Procedure: Using a thermometer, measure the temperature of the soil suspension. Set this temperature (°C) on the pH meter. Carefully rinse the electrode with DI water and blot dry with tissue paper. Place the electrode in the sample suspension to check the nature (alkaline or acidic) of the soil. Take out the electrode, rinse with DI water and blot dry with a soft tissue.

According to the samples nature, calibrate the pH meter using pH 4.0, 7.0 and 10.0 buffer solutions. If the soil sample is basic in nature, the pH meter must be calibrated using pH 7.0 and 10.0 buffer solutions. On the other hand, if the soil sample is acidic in nature, the meter must be calibrated using the pH 4.00 and 7.00 buffer solutions. Take the electrode out from the buffer solution, rinse with DI water, and blot dry with a soft tissue. Then place the electrode in the sample soil suspension. Record the pH of the soil sample in both water and 0.01 M $CaCl_2$ aqueous solution. Turn off the pH meter.

Results and discussion: The pH measured using aqueous and $CaCl_2$ solution can be expressed as pH (w) and pH($CaCl_2$), respectively. As the concentration of salts varies with the moisture content of the soil, the effect of these variations on pH can be minimized using 0.01 M $CaCl_2$. Hence, the pH test using $CaCl_2$ solution gives more accurate results. The values of pH(w) are usually higher than pH($CaCl_2$) by 0.5 to 0.9 unit.

2.5 COMPLEXOMETRIC TITRATION

Metal ions can be quantitatively estimated by complexometric titrations using a chelating ligand or macrocyclic ligand like ethylenediaminetetraacetic acid (EDTA). Complexometry is also familiar as chelatometry. By using suitable indicator (also called metallochromic indicators) one can easily perform the titration and calculate the exact amount of a desired metal ion concentration in a soil sample.

In complexometry, EDTA is the most popular chelating agent due to its unique characteristics like low toxicity, high stability or formation constant (K_f), and excellent selectivity within wide range of pH and masking agents. It reacts with alkaline earth metal ions and transition metal ions in 1:1 stoichiometry. The complexation of metal ions by EDTA (Y^{4-}) is represented in Fig. 2.4.

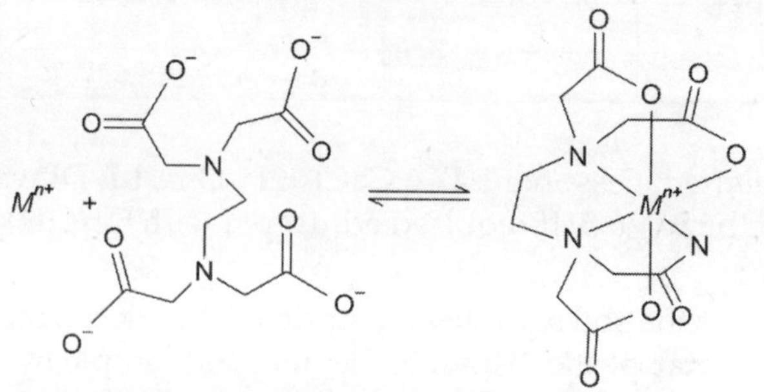

Fig. 2.4: Complexation of metal ion by EDTA

Chelation: When the central metal ion in a complex is bound to two or more donor atoms of one ligand, the complex is called *chelate* and the process of formation of a chelate is called *chelation*. Chelation gives exceptional stability to the complex.

The examples of a chelate complex include $[Cu(en)_2]^{2+}$, $[Co(en)_3]^{3+}$, $[Ni(dmg)_2]^{2+}$ and heme unit (Fe(II)-porphyrinate complex) of haemoglobin.

Chelating agents: If a multidentate ligand is able to coordinate to a metal ion by using two or more donor atoms simultaneously by donating two or more pairs of electrons thus forming a chelate complex, is called a *chelating agent*.

The complexometric titration, metallochromic indicators and buffer systems for maintaining the pH of reaction medium are discussed in details in Chapter 7.

2.6 ESTIMATION OF CALCIUM AND MAGNESIUM IONS AS CALCIUM CARBONATE BY COMPLEXOMETRIC TITRATION

Experiment 2.2

Aim: Determination of Ca^{2+} and Mg^{2+} ions concentrations in soil as $CaCO_3$ by complexometric titration method.

Principle: The titrimetric method for determination of Ca^{2+} and Mg^{2+} ions using EDTA titrant is developed by Chang and Bray (1951). This method is based on the principle that EDTA forms stable complexes with Ca^{2+} and Mg^{2+} ions at different pH. The interference from heavier elements like Cu, Zn, Fe, Mn can be eliminated by using 2% NaCN solution or carbamate. As the concentration of interfering ions is negligible in irrigation waters and water extracts of soil, it can be ignored.

To determine the amounts of Ca^{2+} and Mg^{2+} ions present in a soil extract solution, firstly a portion of the extract is titrated against standard EDTA using murexide indicator and carbamate as masking agent at pH 12, which gives only the amount of Ca^{2+} ions present in that mixture. Then another portion of the mixture solution is again titrated with standard EDTA solution using EBT indicator at pH 10, which gives the total concentration of Ca^{2+} and Mg^{2+} ions present in a soil sample. From the difference, the concentration of Mg^{2+} ions present in the sample is calculated.

Apparatus: Shaker, porcelain dish, beakers, volumetric/conical flask.

Chemical reagents:

  (i) *Standard 0.01 N Ca^{2+} solution:* Weigh 0.5 g of pure $CaCO_3$ in 400 mL beaker and add 10 mL of 3N HCl. Boil the solution on a water-bath to expel CO_2. After the effervescence stops filter the solution in a 1 L volumetric flask and then dilute the solution up to the mark with DI water.

 (ii) *NH_3–NH_4Cl buffer solution:* Dissolve 17.38 g of NH_4Cl in 143 mL of conc. NH_4OH (18.1 M, 35%) in a 400 mL beaker under fume-hood. Dilute to 250 mL with DI water, homogenize and store in an amber glass flask in the refrigerator.

(iii) *4 N NaOH solution:* Dissolve 160 g of pure NaOH in DI water and make the volume to 1 L. This will give pH 12.

(iv) *0.01 M EDTA solution*: Take 3.72 g of ethylenediaminetetraacetic acid disodium salt dihydrate in a 400 mL beaker and add 200 mL DI water. After making a homogeneous solution transfer it to a 1000 mL graduated cylinder and dilute it with DI water up to the mark. Then keep the solution in a polyethylene flask.

(v) *Solid muroxide indicator:* Weigh 0.5 g of murexide and 49.5 g of NaCl. Mix the two solids and grind the mixture well. Finally store the mixture in an opaque flask.

(vi) *EBT indicator:* Dissolve 1 g of EBT in 100 mL 95% ethanol.

(vii) Sodium diethyl dithiocarbamate crystals

Sample preparation: Weigh 5 g air dried soil sample in 250 mL beaker and add 25 mL neutral NH_4OAc solution. Shake the mixture using a mechanical shaker for 5 minutes. Filter the mixture solution through Whatman No.1 filter paper and collect the filtrate for analysis.

Procedure:

(i) *Standardization of 0.01 M EDTA solution:* Pipette out 10 mL of standard Ca^{2+} solution in a 250 mL conical flask and dilute it to 50 mL using DI water. Then add 5 mL 4 N NaOH solution followed by a pinch of murexide indicator. Titrate the solution against 0.01 M EDTA until the colour changes from light pink to violet colour.

(ii) *Determination of Ca^{2+} only:* Pipette out 10 mL aliquot of the soil extract and add 2–3 crystals of carbamate followed by addition of 5 mL of 4 N NaOH solution. Then add a pinch of murexide indicator powder. Titrate the solution against standard 0.01 M EDTA solution until the colour changes from pink to violet. Perform the titration in triplicate. The average volume of the EDTA solution required in this step is noted by V_1 mL.

(iii) *Determination of total Ca^{2+} and Mg^{2+} ions:* Using a graduated pipette take out 10 mL aliquot of the soil extract in a 250 mL conical flask and dilute to 50 mL with DI water. Then add 2–3 crystals of carbamate, 5 mL of NH_3–NH_4Cl buffer solution and 3–4 drops of EBT indicator. Titrate the solution against standard EDTA solution until the colour changes from wine-red to blue. Perform the titration in triplicate. The average volume of the EDTA solution required in this step is noted by V_2 mL.

Calculation: Volume of EDTA required for $Ca^{2+} = V_1$ mL

Volume of EDTA required for $Mg^{2+} = (V_2 - V_1)$ mL

Amount of CaO and ZnO can be calculated from the equations below:

1 mL of 1 M EDTA $\equiv$ 0.05608 g of CaO

1 mL of 1 M EDTA $\equiv$ 0.08138 g of ZnO

1 mL of 1 M EDTA $\equiv$ 0.10008 g of $CaCO_3$

Alternative Method

Aim: Determination of total Ca^{2+} and Mg^{2+} ions in soil as $CaCO_3$ by complexometric titration method.

Principle: The total concentration of Ca^{2+} and Mg^{2+} ions in soil sample can be estimated by complexometric titration using EDTA and eriochrome black T (EBT) indicator. The method used here is back titration. For the titration, the soil extraction containing both Ca^{2+} and Mg^{2+} ions is treated with an excess of EDTA. At this stage, the indicator is added. As all the ions are complexed with EDTA, the indicator remains in free form resulting in a blue colour of the solution. Then the excess EDTA is back titrated using a standard solution of $MgCl_2$. The added Mg^{2+} ions from the burette form complex with the excess EDTA molecules. At the end point, when all the excess EDTA is complexed, the Mg^{2+} ions form complex with the indicators immediately changing the colour from blue to pink.

The main reaction:
$$Ca^{2+} + Mg^{2+} + EDTA^{4-} \rightarrow [Ca(EDTA)]^{2-} + [Mg(EDTA)]^{2-}$$

Back titration:
$$EDTA^{4-} + Mg^{2+} \rightarrow [Mg(EDTA)]^{2-}$$

Indicator reaction:
$$EBT + Mg^{2+} \rightarrow [Mg(EBT)]$$
$$\text{(blue)} \qquad\qquad \text{(pink)}$$

Apparatus: Beaker, glass rod, volumetric flask, filter paper, conical flask, pipette, burette, glass rod, filter paper, pH paper.

Reagent:

(i) Dil. HCl (2 M)

(ii) Dil. NaOH solution (2 M)

(iii) *NH_3–NH_4Cl buffer solution* (pH 10): Dissolve 17.38 g of NH_4Cl in 143 mL of conc. NH_4OH (18.1 M, 35%) in a 400 mL beaker under fume-hood. Dilute to 250 mL with DI water, homogenize and store in an amber glass flask in the refrigerator.

(iv) *1% (w/v) EBT indicator:* Dissolve 1 g of EBT in 100 mL 95% ethanol.

(v) *0.05 M EDTA solution:* Weigh 9.31 g of the disodium EDTA salt in a 500 mL volumetric flask and add 50 mL DI water. Dissolve the salt completely and dilute the solution up to the mark with DI water. Make it homogeneous.

(vi) *0.05 M $MgCl_2$·$6H_2O$:* Weigh 5.08 g of $MgCl_2$·$6H_2O$ in a 500 mL volumetric flask. Dissolve in 50 mL DI water. Dilute the solution up to the mark with DI water and homogenize.

Sample preparation: Weigh ~0.5 g of the soil sample into a 100 mL beaker and add 20 mL dil. HCl. Stir the solution with a glass rod and allow the solid to dissolve completely. Neutralise the excess acid with dil. NaOH solution until pH 7 (check with pH paper). Transfer the solution quantitatively to a 100 mL volumetric flask via filtration and dilute it up to the mark with DI water.

Procedure:

Determination of total Ca^{2+} and Mg^{2+}: Pipette out 10 mL of the soil extract solution into a 250 mL conical flask and add 20 mL of 0.05 M EDTA solution. Then add 10 mL of ammonia buffer solution. Dilute the solution with 50 mL DI water followed by addition of 1 mL of EBT indicator solution. Titrate the mixture solution with the standard 0.05 M $MgCl_2$ solutions until the end point is indicated by a colour change from blue to pink.

Standardization of EDTA: Pipette out 10 mL aliquot of the standard magnesium sulphate solution. Then add 2–3 mL of NH_3–NH_4Cl buffer solution. Then add a pinch of solid EBT indicator or 1 drop of EBT indicator. Titrate the solution against 0.01 M EDTA until the colour changes from wine-red to blue colour.

Calculation: Volume of EDTA initially used (V_1) = 20 mL

Volume of Ca^{2+} required to back titrate EDTA = V_2 mL

Amount of total Ca^{2+} and Mg^{2+} ions can be calculated from difference of the two volumes (V_1 and V_2) using the equation below:

1 mL volume difference of 1 M EDTA ≡ 0.10008 g of $CaCO_3$

BIBLIOGRAPHY

1. Kalra YP. Determination of pH of soils by different methods: Collaborative study. *Journal of AOAC International*, 1995, 78(2), 310–24.
2. Soil Testing in India, Methods Manual, Department of Agriculture & Cooperation Ministry of Agriculture, Government of India, New Delhi, January, 2011.
3. Vogel AI and Jeffery GH. *Vogel's Textbook of Quantitative Chemical Analysis*. Wiley, 1989.

QUESTIONS

Multiple Choice Questions

1. What is the approximate percentage of inorganic minerals in soil?
   (a) 30%   (b) 45%
   (c) 5%   (d) 25%
2. The smallest particle in solid phase of soil is:
   (a) Sand   (b) Stone
   (c) Clay   (d) Silt
3. Match the particles of soil (Column I) with solid components (Column II).
   Column I   *Column II*
   A. <0.002 mm   (i) Coarse sand particle
   B. 0.002 to 0.02 mm   (ii) Fine sand particle
   C. 0.02 to 0.2 mm   (iii) Silt
   D. 0.2 to 2.0 mm   (iv) Clay
   (a) A – iv, B – iii, C – ii, D – i   (b) A – iv, B – iii, C – i, D – ii
   (c) A – iii, B – iv, C – ii, D – i   (d) A – i, B – ii, C – iii, D – iv
4. What is the role of organic matter in soil?
   (a) It improves water infiltration
   (b) It breaks down organic pollutants
   (c) It converts N_2 in the air into nitrates used by plants
   (d) It is rich in nutrients, which is important for fertility
5. pH is defined as:
   (a) $-\log[H^-]$ (b) $-\log[H^+]$
   (c) $+\ln[H^+]$ (d) $+\ln[H^-]$
6. Which pH range of soil is ideal for plant growth?
   (a) 3–5   (b) 5–8
   (c) 9–12   (d) None of these
7. Below which pH plant growth is not possible?
   (a) 5.4   (b) 6.2
   (c) 3.7   (d) 4.9
8. At which pH range most of the soil nutrients become available to plants and crops?
   (a) 7.6–8.3   (b) 5.5–6.5
   (c) 4.5–5.2   (d) >7.4
9. The preferred pH range for colonization of rhizobia is:
   (a) 3.4–4.6   (b) 5.0–7.0
   (c) 7.4–8.6   (d) 6.0–7.5

10. In complexometric titration, EDTA acts as a .... ligand.
    (a) Hexadentate   (b) Tetradentate
    (c) Bidentate   (d) Monodentate
11. Which of the following is a metallochromic indicator?
    (a) Phenolphthalein   (b) Starch
    (c) EBT   (d) None of these

Answers

1. (b); 2. (c); 3. (a); 4. (d); 5. (b); 6. (b); 7. (c); 8. (b); 9. (d); 10. (a); 11. (c)

Practice Questions

1. What is soil? Discuss its major constituents.
2. Write down the relative percentage of minerals, organic matter, water and air in soil.
3. Write down the importance of soil analysis.
4. What do you mean by soil pH. How it regulates the nature of soil?
5. Discuss briefly the effect of pH on the available nutrients present in soil.
6. How pH is related to plant growth on a particular soil.
7. How the soil pH affects the colour of *Hydrangea* flowers?
8. Discuss briefly the procedure for determination of soil pH.
9. Write a short note on complexometric titration.
10. What do you mean by metallochromic indicator? Give one example.
11. What is chelation? Define chelating agent with an example.
12. Write the principle for the estimation of calcium and magnesium in soil by complexometric titration.

Chapter

3

Analysis of Water

UGC Syllabus

Definition of pure water, sources responsible for contaminating water, water sampling methods, water purification methods.
1. Determination of pH, acidity and alkalinity of a water sample.
2. Determination of dissolved oxygen (DO) of a water sample.

INTRODUCTION

Water is the most essential element for existence of life on earth and it is present in sufficient quantity as natural sources. About 75% of total water content covers the earth's surface in the form of snow over mountains and as liquid water in springs, rivers, lakes, oceans, etc. Out of this 75%, only 1% is present as freshwater which is available for human consumption.

Due to various human activities like agriculture, industrial, mining, etc. water is getting polluted day-by-day. Therefore, it is very essential to purify waters from different sources before its utilisation in various proces-ses. Proper analysis of water is a must prior to the purification process.

3.1 CONCEPT OF PURE WATER

The denotation of the term 'pure water' depends on the individual's perspective as it does not exist in nature. Being the universal solvent, water generally dissolves all the particles and minerals that come in its path. When raindrops fall to earth, they engulf almost all the mine-rals in the atmosphere and as soon as they reach the ground they catch minerals from the soil, finally ending up as groundwater and other water bodies. Hence, most of the water contains certain ions, such as Ca^{2+} and Mg^{2+} even if it may be in just a trace amount. These ions determine the nature of water (hard or soft) and play important role in the taste of the water.

From the perspective of general people 'pure water' for drinking signifies to bacteria free water not the chemical free water. Water purifier companies produce healthy water by recognizing the unhealthy impurities in the water and then removing them by proper action. Thus, the public discussion on water switches from 'pure' to 'healthy' water which has a suitable pH range (pH = 7 to 8) for human biological processes. On the other hand, from a scientific perspective water purity is defined by its expected use. The water which is required to be used in industry, agriculture, or laboratory should be 'fit for use'.

Definition of pure water: The water after removal of all kinds of impurities such as mineral and organic matter is called *pure water* or *purified water*. Water purification can be performed in many ways like microporous filtration, carbon filtration, and ultraviolet oxidation. The most commonly used pure water for laboratory work is of two types—distilled water and deionized water.

- *Distilled water:* This type of water is purified by distillation. The process involves heating of water to its boiling point producing steam where the impurities are left behind. Then the steam is condensed into a sanitary container and the condensate is called distilled water. This process virtually removes all inorganic and organic contaminants including microorganisms.

- *Deionized water:* This type of pure water is obtained by ion-exchange method. Deionisation is a physical process that uses ion exchange resins which exchange H^+ and OH^- ions for the dissolved ions which then recombine to form water. In this method all mineral ions such as calcium, sodium, iron, copper, chloride and sulphate are removed from the water. However, this process is not so efficient to eliminate uncharged organic molecules or microorganisms.

Compared to distillation, deionization is less expensive and cheaper way to purify water. However, it is not capable of producing water with the same purity and consistency of distillation. Hence, distilled water may be used in replacement of deionized water but the reverse is not always possible.

3.2 WATER CONTAMINATION

Sources of Water Contamination

Three major sources of water are rainwater, surface water and groundwater. All of these three water systems are continuously getting contaminated from various sources (Fig. 3.1). These sources are discussed below.

- *Rainwater contamination:* Rainwater is comparatively free from impurities than other sources. However, with increasing air pollution, it gradually gets contaminated. Wind-blown dirt or dust, smoke, soot, leaves, birds poop, etc. are the sources of contamination for rainwater during rainfall. It can also carry pathogenic microorganisms and chemicals. So, the quality of rainwater becomes unhealthy during harvesting and storing.

- *Surface water contamination:* Surface water includes various water bodies on earth's surface, i.e. ponds, lakes, rivers, oceans, etc. These water bodies get contaminated by the growing pollution from urban, agricultural and industrial sources. These sources contain heavy metals (mainly Cr^{6+}, Pb^{2+}, As^{3+}), pathogenic microorganisms (virus and bacteria) and organic chemicals (mainly dyes) in very high concentration. The excessive richness of nutrients, more commonly nitrogen and phosphorus, in a water body, stimulates the growth of algae which disrupts normal ecosystem functioning by causing oxygen depletion. This process governed by nutrient pollutant is called eutrophication.

- *Groundwater contamination:* Water from underground is usually considered as safe water to drink, but over the past few decade contamination of the groundwater has increased. Water from deeper aquifers has less contamination. Geology as well as the soil makeup also influences the quality of groundwater. The major

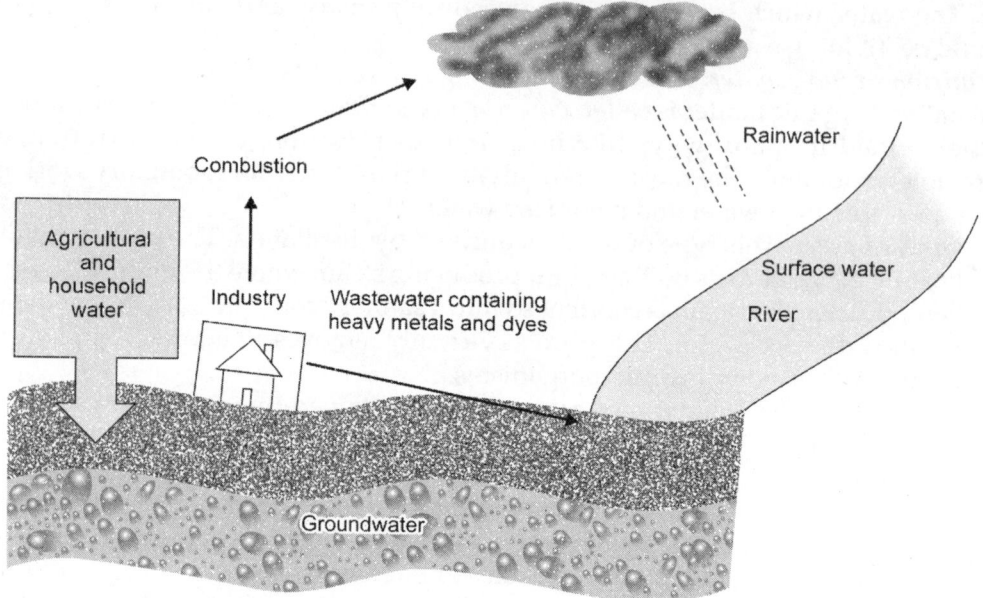

Fig. 3.1: Different sources of contamination of water

contaminants present in groundwater are certain naturally occurring (geogenic) elements like arsenic, selenium or boron. Other sources for contamination of groundwater include sewage, fertilizers, pesticides, on-site sanitation systems, industrial leaks, hydraulic fracturing and landfill leachate.

Categories of Water Contaminants

Different water contaminants are classified into four basic categories:

- *Dissolved solids:* Inorganic minerals containing calcium, magnesium, aluminum ions and other heavy metals like chromium, manganese, iron, copper, lead, cadmium, mercury, arsenic, radon and barium are included in this category.
- *Chemical:* Volatile organic compounds, inorganic chemicals such as chlorine compounds, chloramines, nitrates and various pesticides, herbicides etc. fall under this category.
- *Biological:* The biological contaminants are the pathogens having serious or deadly health effect, including cysts and parasites (such as *Giardia, Cryptosporidium* and tapeworms), bacteria (such as *E. coli, Salmonella, Shigella* and *Legionella*), and viruses (such as hepatitis A and poliovirus).
- *Aesthetic:* The aesthetic contaminant is referred to the invasive tastes and odours that results from various sources. Sediments, dirt, sand or particulates and dissolved minerals can affect the taste and create nuisance by producing foul smell to the water in which they are present.

3.3 WATER SAMPLING METHODS

Sampling is the process of collecting a small but sufficient portion of a material for laboratory analysis. The volume should be small enough for easy transportation and handling, at the same time it should accurately represent the whole system or material. Sampling is the preliminary and most important part of the analytical process. It is so important that in some processes it becomes the main reason for the erroneous results in the analytic process, especially when a trace contamination is being analysed.

General guidelines/principles for water sampling:

(I) *Choice of suitable containers:* While sampling of water, it is very crucial to choose proper container for storage of the sample. Depending on the type of analysis to be carried out, different kind of containers is used. While choosing an appropriate container it should be remembered that the container and stopper should not contaminate the sample (e.g. borosilicate or sodalime glass containers may increase the concentration of silica or sodium in the sample) or absorb/adsorb the constituents (e.g. hydrocarbons, if present in sample, may be absorbed in a polyethylene container) or react with certain component of the sample (e.g. fluorides react with glass). The suitable type of containers for different type of water analysis experiments are listed in Table 3.1.

Table 3.1: Container types and preservatives needed for sampling

Analysis	Container	Minimum volume	Preservation	Maximum recommended preservation time
General	Glass, PE	1000 mL	4 °C, dark	–
Acidity	Glass, PE	100 mL	4 °C, dark	24 h
Alkalinity	Glass, PE	1000 mL	4 °C, dark	24 h
BOD	Glass, PE	1000 mL	4 °C, dark	24 h
COD	Glass, PE	100 mL	H_2SO_4, pH <2	ASAP
DO	BOD bottle	300 mL	DO fixing chemicals	ASAP
Fluoride	PE	300 mL	None	7 days
Phosphate	Glass	100 mL	H_2SO_4, pH <2	–
Pesticides	Glass, teflon	1000 mL	4 °C, dark	7 days
Toxic metals	Glass, PE	500 mL	HNO_3, pH <2	–

Source: Guidelines for water quality monitoring, CPCB (MINARS/27/2007–08)

(II) *Cleaning of containers:* After selecting the suitable containers, it is properly washed with suitable solvents to minimize the probable contamination. For general chemical analysis, a new glass container is thoroughly washed with soap water followed by chromic acid solution ($K_2Cr_2O_7$ + Conc. H_2SO_4) and finally with distilled water to remove the dust and packaging material. However, for the analysis of phosphorous, sample container should not be washed with detergent. Similarly, for the analysis of sulphate and chromium, chromic acid solutions should not be used for washing the container.

Polyethylene containers are cleaned with 1 M HNO_3 or HCl followed by distilled or DI water.

For determination of pesticides, herbicides, and their residues new brown glass containers (amber) should be used. These containers are first cleaned using soap water then with distilled water and finally with hexane or petroleum ether. Then these containers are dried (uncapped) in an oven at 360 °C for 1 h.

For microbiological analysis, the new glass containers are cleaned with non-ionic detergent solution followed by nitric acid with thorough rinsing with distilled water. Then these containers should be sterilized at around 160 °C and at this temperature, they should not produce any chemicals influencing biological activity. For polycarbonate and heat resistant polypropylene containers, a lower sterilization temperature is used. To eliminate bacterial inhibition by chlorine, a 0.1 mL 10% (w/w) sodium thiosulfate ($Na_2S_2O_3$) is added for every 125 mL capacity of container before sterilization.

(III) *Volume of sample:* For most physical and chemical analysis generally two litre water samples are sufficient. However, the quantity may vary depending on the type of analysis.

(IV) *Filling the containers:* After choosing the proper container and volume, one can collect the sample.

(V) *Sample preservation:* Proper preservation of the samples is necessary to get accurate and less erroneous results from the analysis. The preservation conditions of some analyses are listed in Table 3.1.

3.4 WATER PURIFICATION METHODS

Water purification is a method of eliminating unwanted industrial chemicals, biological microorganisms and toxic gases from water.

A large volume of water is purified and disinfected for drinking purpose; however the water is also purified for other applications, such as, in pharmacological, medical, chemical and industrial applications according to their requirements. There are various methods available for water purification including physical (sedimentation, filtration, and distillation), chemical (chlorination and flocculation), biological (activated carbon filters or slow sand filters) processes and sometimes electromagnetic radiation (UV light).

3.4.1 Boiling

Boiling is the oldest and cheapest method for water purification. This is also safest and most commonly practiced household water treatment method. It is extremely effective in removing microbial contaminants from the water. According to WHO, water should be heated for 1 minute with constant appearing of big bubbles on the water surface. However, different microorganisms have the different heat sensitivity. So, sometimes the pure water can be achieved by heating at 70 °C for 15 minutes.

Disadvantages

- It may change the taste of water which becomes unpleasant.
- This process cannot remove chemical toxins or impurities.
- It is not so effective against potentially harmful chemicals or metals.
- Heating method includes fire, stove or kettle, which has the safety issues.
- The boiled water again gets contaminated once it is cooled.

3.4.2 Filtration

This is a low cost and effective way to purify water using the right filter. This process removes different kind of chemicals and numerous other dangerous pollutants. Filtration is much effective method than others, as the purified water using this process still has some minerals. This method involves chemical absorption mechanism that effectively eliminates undesired substances from water.

Disadvantages
- This process cannot eliminate all germs and contaminants.
- Using proper filter is very essential. Otherwise, very small particles can pass through the membranes.
- Careful cleaning and handling of the equipment is very important.
- Disposal of cartridges is also important, as it contains harmful toxins which are potential threats to the environment.

3.4.3 Distillation

The process involves heating of water to its boiling point producing steam where the impurities are left behind. Then the steam is condensed into a sanitary container and the condensate is called distilled water. This process virtually removes all inorganic (inorganic salts and heavy metals) and organic contaminants including microorganisms. It is assumed that the water has lower boiling point than the contaminants. So that the water is vaporised and the other high boiling substances are deposited as sediments.

Disadvantage
- It is very slow process.
- Some pesticides and herbicides which have lower boiling point than water are also condensed along with water vapours.
- It needs a heat source which has the safety issues. It is not ideal for large scale production such as in commercial or industrial purpose.

3.4.4 Chlorination

Chlorination is a very old chemical method by which drinking water is disinfected with addition of chlorine or chlorine-containing substances such as $NaOCl$ solution or solid $Ca(OCl)_2$. It is a high scalable, ease-of-use and low cost method. Chlorine destroys pathogens and other organisms by destroying the cell membrane and DNA activity.

Disadvantages
- The taste and smell of the chlorinated water is different.
- It gives relatively low protection against protozoa.
- Lower disinfection effectiveness is achieved in turbid waters.
- A safe level (up to 4 mg/L or 4 ppm) of chlorine should be used as a high level of chlorine could be poisonous in drinking water.
- Chlorine reacts with naturally occurring organic compounds, forming disinfection by-products (DBPs) such as, haloacetic acids (HAAs) and trihalomethanes (THMs) which causes serious health hazards.

3.4.5 Reverse Osmosis

Reverse osmosis (RO) is a water purification process that eliminates the impurities of the feed water when pressure forces it through a semipermeable membrane (Fig. 3.2). The amount of pressure needed depends on the concentration of the salt in feed water. For highly concentrated feed water, the high pressure is needed to overcome the osmotic pressure. When the water pressure drives water across the RO membrane and additional filters, such as sediment or carbon filters, the contaminants are filtered out and subsequently flushed down the **drain**.

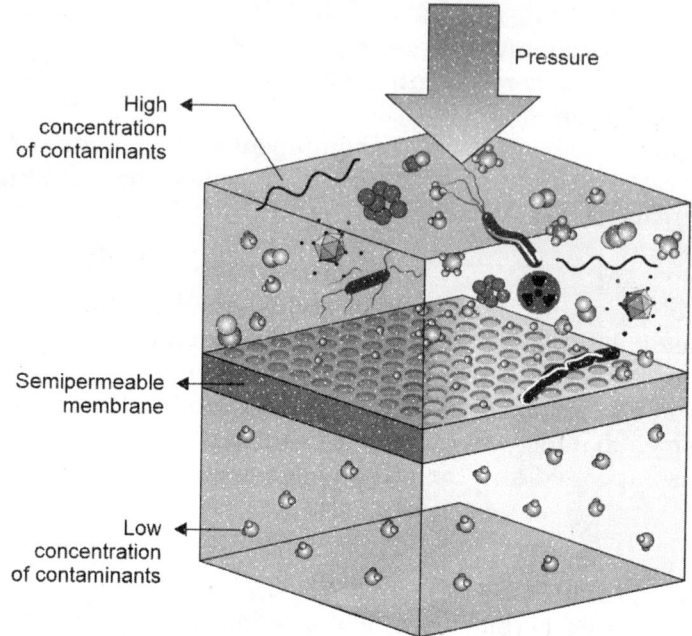

Fig. 3.2: Schematic representation of reverse osmosis process

This process effectively eliminates ~99% of the dissolved salts (ions), colloids, organics, particles, bacteria and pyrogens. A RO membrane works on the basis of the molecular size and ionic charge of the impurities. Any impurity having molecular weight greater than 200 Da is not allowed to pass through the RO membrane. Likewise, a small ionic charge element easily passes through the RO membrane than the element having high ionic charge. With the same reason the gases, i.e. non-ionic species like CO_2 are not rejected by the RO membrane. The removal of salt from seawater (desalination) is also done by reverse osmosis.

Disadvantages
- The main disadvantage of this process is wastage of large amount of water.
- This is an expensive process with slow output.
- It does not disinfect water to make it pure and clear.
- Periodic replacement of the RO membranes is very important. The lifespan of the RO membrane is comparatively short for hard water.
- This method is not suitable for the water which contains harmful microorganisms.
- A lot of energy is needed for the entire process.

3.4.6 Slow Sand Filtration

Slow sand filtration (SSF) is another cost-effective method to eliminate the protozoa and most of the bacteria which causes gastrointestinal diseases. Feed water passes through a layer of sand where it is not only physically filtered, but also biologically treated. Both the sediments and pathogens are eliminated. This method is based on the ability of organisms to eliminate pathogens. Few layers of sand and gravel are kept in a concrete or plastic container. Above this sand layer 5–6 cm of the container is filled with water allowing the growth of a biofilm (bioactive layer) on the top. This film helps to remove the microorganisms that causes various diseases. During addition of water a diffuser plate is used to protect the biolayer. This process is one-time installation process which has long life with no recurrent expenses.

Disadvantages

- This process has low filtration rate.
- It is not so effective against viruses.
- Routine cleaning can harm the bio-layer and decrease effectiveness.
- Feed water turbidity must generally be low with low algae contents.
- This method has no chlorine residual protection which can lead to recontamination.

3.5 pH OF WATER

To begin with, pH is a scale to compare the acidity or alkalinity of different aqueous solutions. In the year of 1887, Arrhenius put forward his theory of ionisation. According to Arrhenius concept, acids and bases are those substances which produce H^+ and OH^- ions, respectively, in solution (aqueous). Since pure water ionizes to produce equivalent amount of H^+ and OH^- ions, acids are supposed to increase H^+ ion concentration in aqueous solution whereas bases increase amount of OH^- ions. Again different acids and bases have different degree of ionisation. So, the hydrogen ion concentration of all acid solutions or hydroxyl ion concentrations of all base solutions are not same. In 1909, Sørensen introduced the term pH as a measure of hydrogen ion concentration of aqueous solutions.

The pure water yields equal concentration of hydrogen ions and hydroxyl ions $(10^{-7}\,mol/L)$.

$$H_2O \rightleftharpoons H^+ + OH^-$$

From the law of mass action, it can be shown that, for pure water at about 25°C:

$$[H^+] \times [OH^-] = K_w \text{ (ionization constant for water)} = 10^{-14}$$

Sørensen defined pH of a solution as the negative logerithm of the H^+ ion concentration:

$$pH = -\log[H^+]$$

The pH scale ranges from 0 to 14, with pH 7 at 25 °C representing neutrality (pure water). A pH <7 represents acidic conditions of the solution whereas alkaline (base) conditions produce a value of pH >7. The hydrogen ion concentration may be measured by the pH meters via a glass electrode and a saturated calomel reference electrode. The pH of water is measured in order to calculate dissolved carbonate, bicarbonate and CO_2 concentration, corrosion and stability index, etc.

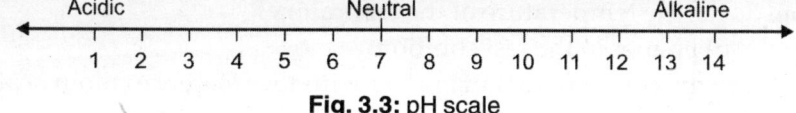

Fig. 3.3: pH scale

Experiment 3.1

Aim: Determination of water pH (i) using pH paper (colorimetrically) and (ii) using pH meter (electrometrically).

Principle: The first method (colorimetrically) of pH measurement is used for rough estimation. In this method a universal indicator or pH paper is used. The acidity or alkalinity is confirmed by merely the colour change of the indicator used. On the other hand if pH paper is used it gives the value of pH of a solution as an integral value not accurate up to first decimal place.

The electrometric method is therefore used to get the exact value of pH of a solution. A glass electrode is used as working electrode along with a saturated calomel reference electrode. Here, activity of hydrogen ions is obtained potentiometrically. The glass electrode is brought in contact with the test solution by using a liquid junction. The electromotive force (EMF) of the system is measured using a pH meter which is a high impedance voltmeter calibrated in terms of pH. The glass electrode generates a potential varying linearly with the pH of the solution in which it is immersed. The system is so designed that a change of 59.1 mV electric charge changes the pH by 1 unit at 25 °C.

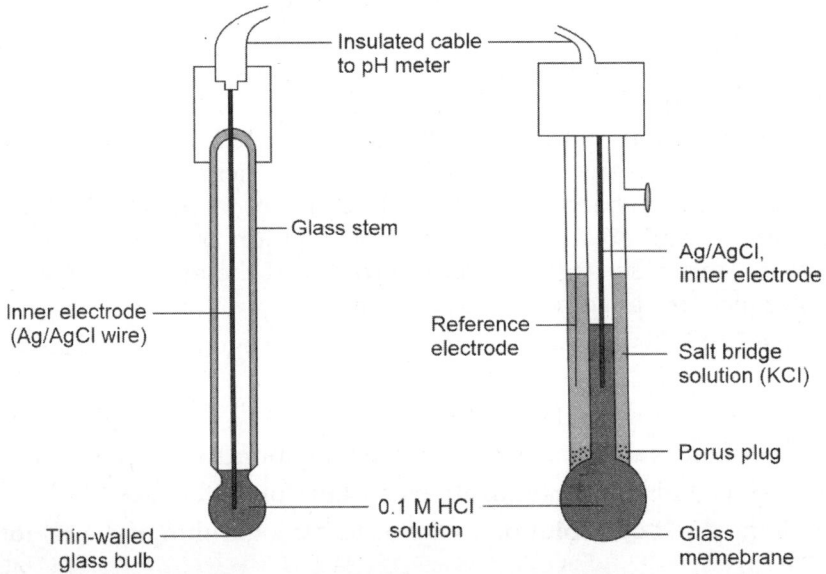

Fig. 3.4: Common pH electrodes; glass electrode (left) and combination electrode (right)

Apparatus:

  (I) *Colorimetric method:* Wide range pH paper

  (II) *Electrometric method:* A pH meter, glass electrode, and Ag/AgCl/KCl or a calomel reference electrode.

Procedure:

  (I) *Colorimetric method:*

      (i) Note down the temperature of the laboratory.

      (ii) Dip the pH paper in the test solution.

      (iii) Compare the colour of that pH paper with the reference colour scale.

      (iv) Record the value of pH of the sample.

  (II) *Electrometric method:*

      (i) Dip the pH electrode in a buffer solution of known pH. Standardise the pH meter using calibrating knob.

(ii) After calibrating the pH meter, wash the electrode properly with distilled water to get rid of any adhered buffer. Pat dry the electrode using tissue paper.

(iii) Take the test solution in a beaker and adjust the temperature knob so that the solution temperature and instrument temperature becomes same.

(iv) Dip the electrode in the test solution taken in a beaker. Take the reading.

(v) Wash the electrode again by dipping in distilled water and measure the pH of the test solution two more times.

Observation:

Water sample	pH			Average pH
	1st dip	2nd dip	3rd dip	
A				
B				
C				

Result: The pH of the given water samples A, B and C are found to be ....., ... and ....., respectively, which indicates that the samples are acidic/alkaline/neutral in nature.

Precautions:

(i) The glass bulb of the pH electrode should always be clean, to avoid incorrect result.

(ii) Stock solution is collected immediately before pH estimation or the value changes.

Environmental significance of pH: The determination of the pH of water is very necessary, as it plays an important role in determining the nature of the water. For drinking water, the pH must be within the range of 7.0 to 8.5. A high pH value of water (alkaline water) results in scale formation in the apparatus when the water is heated in it. Alkaline water also decreases the germicidal potential of chlorinated chemicals. On the other hand at pH below 6.5 (acidic water) the water becomes corrosive to the pipes and releases toxic metals from it.

The value of water pH must be known before treating the water in various purification techniques like disinfection, coagulation (suitable pH range for alumn treatment: 6.5–8.5), corrosion control, bio-treatment (favourable pH: 5–10) and water softening. Efficiency of all these processes is very much sensitive to the pH value of water.

Hence, the pH determination followed by required treatment of the water to adjust the pH is very essential for proper selection of disinfectant and chemical coagulant.

3.6 ACIDITY OF WATER

The ability of water to neutralise bases is known as acidity of water. Pure water is neutral (pH = 7) in nature. Contaminated water may show acidic behaviour due to dissolved weak organic acids such as tannic acid or acetic acid and strong mineral acids like HCl, H_2SO_4, etc. On the other hand, dissolved CO_2 is the main source of acidity in unpolluted water. Almost all natural waters contain some carbon dioxide. Groundwater and hypolimnion waters from different water bodies often contain considerable amounts of CO_2 resulting from both aerobic and anaerobic oxidation of organic contaminants by bacteria. However, when the CO_2 concentration of surface water is less than the CO_2 concentration of the atmosphere remaining in close proximity of the water surface, the surface water absorbs CO_2 (Henry's law). Industrial wastewaters from metallurgical and organic synthetic industries contain mineral acidity.

Experiment 3.2

Aim: Determination of the acidity of water sample.

Principle: Acidity of water and wastewater mainly results from dissolved CO_2 (as H_2CO_3), weak organic acids, and mineral acids. Dissociation or hydrolysis of these substances release hydrogen ions in the water which lowers the pH indicating the water sample as acidic one. To measure the amount of acidity water sample is titrated with standard NaOH solutions. The standard methods to determine water acidity involves titration with NaOH to pH 3.7 for *mineral acidity* and to pH 8.3 for determination of *total acidity*. As methyl orange changes colour from red to orange at pH ~3.7, it is employed as indicator for mineral acidity determination and the acidity value calculated from this titration is called *methyl orange acidity*. Titration with phenolphthalein (end point pH 8.3) indicates the neutralization of carbonic acid to bicarbonate and the result of titration is named *phenolphthalein acidity*. As the dissolved CO_2 is the main factor that leads to acidity of natural waters, *phenolphthalein acidity* is often termed as *total acidity*. However, the terms *methyl orange acidity* and *phenolphthalein acidity* are only used to designate the results. Results of the acidity analysis are expressed in mg/L of $CaCO_3$.

Collection and storage of samples: Water samples are collected in polyethylene bottles or borosilicate glass bottles and stored at a low temperature. Fill bottles up to the neck and cap tightly so that no air remains above the sample in the bottle. As wastewater samples are sensitive to microbial action and subject to change in dissolved gas concentrations on air exposure, samples should be analysed without any delay, preferably within 24 h. For biologically active samples analysis should be carried out within 6 h. It is desirable to avoid prolonged air exposure and sample agitation.

Apparatus: Borosilicate glass bottles, burette, burette stands, pipette, conical flasks.

Reagents:
(i) 0.02 N standard NaOH solution
(ii) Methyl orange: pH 3.7 indicator
(iii) Phenolphthalein: pH 8.3 indicator

Procedure:
(I) *Methyl orange acidity:* Take out 25 mL of water sample from the stock and pour in a conical flask. Add 2 to 3 drops of methyl orange indicator to the solution. If the solution becomes yellow in colour, it implies no mineral acidity present in the water sample. The presence of mineral acidity is confirmed when orange colour solution is generated on addition of the indicator. The solution is then titrated with standard 0.02 N NaOH until the end point is confirmed by a yellow colour solution. Record the titrant volume (say, V_1 mL).

(II) *Phenolphthalein acidity or total acidity:* Take a conical flask and pour 25 mL of the sample water. After that, add few drops of the phenolphthalein to it. A pink colour solution indicates no acidity is present in the sample. If solution does not give any colour, acidity (from dissolved weak acids) is present and the sample is titrated against 0.02 N NaOH till the end point is confirmed by a pink colour solution. Record the volume of NaOH solution (say, V_2 mL) consumed in this titration.

Observation:

Experiment	Sample number	Sample volume (mL)	Initial burette reading (V_1 mL)	Final burette reading (V_2 mL)	Volume of NaOH consumed {($V_1 - V_2$) mL}
Methyl orange acidity	A1				
	A2				
	A3				
Phenolphthalein acidity	B1				
	B2				
	B3				

Calculation:

(I) Methyl orange acidity in mg/L as
$$CaCO_3 = (V_1 \times N \times 50 \times 1000)/(\text{Sample volume})$$
(II) Phenolphthalein acidity in mg/L as
$$CaCO_3 = (V_2 \times N \times 50 \times 1000)/(\text{Sample volume})$$
where V_1 and V_2 = mL of NaOH titrant consumed and N = Normality of NaOH

Results:

(I) Methyl orange acidity in mg/L as $CaCO_3$ = ...
(II) Phenolphthalein acidity in mg/L as $CaCO_3$ = ...

Environmental significance: Acidic waters are corrosive in nature. The corrosiveness of most waters results from dissolved CO_2, but in case of industrial wastes it mostly results from mineral acids dissolved in water. The acidity of water largely influences chemical and biological processes. Therefore, prior determination of the acidity value of water is important. Determination of acidity is very essential in case of water supplies to get an idea about the dimension of equipment, storage space, chemicals required and cost of the treatment. The mineral acidity of industrial waste water must be neutralized prior to its discharge in any water body.

3.7 ALKALINITY OF WATER

The ability of water to neutralize acids is known as alkalinity of water. The primary reason for alkalinity of water is the presence of salts of weak acids, and/or weak or strong bases. The major portion of alkalinity of surface waters is caused by hydroxides, bicarbonates and carbonates. Thus, alkalinity is expressed in terms of $CaCO_3$ equivalent to H^+ ions neutralized.

Boiler water is always alkaline as it contains hydroxide and carbonates. High alkalinity of chemically treated water causes precipitate deposition on boiler tubes making the tubes brittle. Calcium and magnesium bicarbonates result in temporary hardness of water. The growth of phytoplankton and algae in natural water is favoured by highly alkaline water.

Experiment 3.3

Aim: Determination of hydroxide, carbonate and bicarbonate alkalinity of water sample.

Principle: The hydrolysis of **salts present** in water results in formation of free hydroxyl ions which is the main reason of alkalinity of water. Thus, the alkalinity of a water sample is determined by the concentration of this hydroxyl ion. Carbonate and bicarbonates also contribute to the total alkalinity. Alkalinity can be measured by titration with 0.02 N H_2SO_4 and is reported in terms of $CaCO_3$ equivalent.

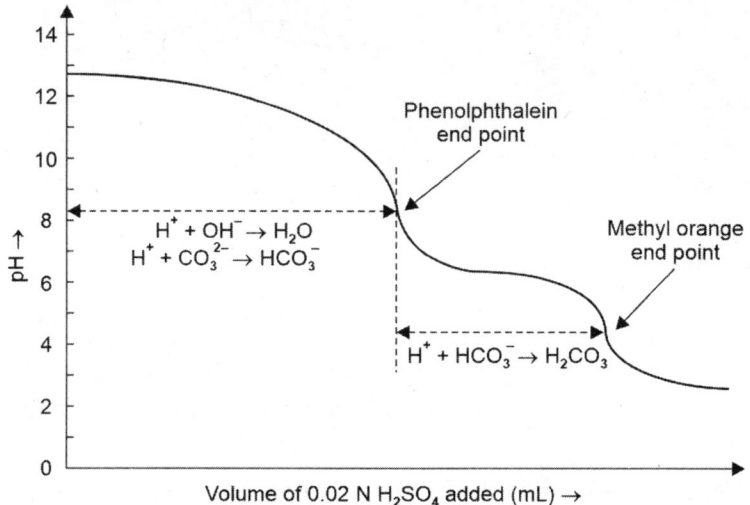

Fig. 3.5: pH metric titration method of alkalinity determination

For samples with initial pH value above 8.3, neutralisation of the alkalinity takes place in two steps. The first titration is carried out in presence of phenolphthalein indicator. A colour change from pink to colourless at pH ~8.2 indicates the end point when full neutralization of hydroxyl ions and half neutralization of carbonate ions are completed.

$$H^+ + OH^- \rightarrow H_2O$$

and
$$CO_3^{2-} + H^+ \rightarrow HCO_3^- \quad (pH = ~8.2)$$

The second titration is conducted in presence of methyl orange indicator. The methyl orange end point (pH = 4.5 to 3.7) corresponds to the full conversion of bicarbonate ion to carbonic acid. The methyl orange end point thus corresponds to *total alkalinity* of the water sample.

$$HCO_3^- + H^+ \rightarrow H_2CO_3 \quad (pH = ~3.7)$$

Collection and storage of samples: Same as described in experiment 3.2.

Apparatus: Burette, pipette, conical flasks, glass bottles.

Reagents:

 i. Standard sulphuric acid (0.02 N)

 ii. Methyl orange indicator (pH 3.7)

 iii. Phenolphthalein indicator (pH 8.3)

Procedure:

(I) *Phenolphthalein alkalinity:* Take 25 mL of water sample in a conical flask and add few drops of phenolphthalein indicator. If pink colouration is observed, titrate the solution with 0.02 N H_2SO_4 in a burette till the colour disappears indicating phenolphthalein end point. Record the volume of H_2SO_4 consumed.

(II) *Methyl orange alkalinity or total alkalinity:* Take 25 mL of water sample in a conical flask and add few drops of methyl orange indicator. If yellow colouration is observed, titrate the solution against 0.02 N H_2SO_4 till an orange colour solution is obtained indicating the end point. Note down the total volume of H_2SO_4 consumed.

Observation:

Experiment	Sample number	Sample volume (mL)	Initial burette reading (V_1 mL)	Final burette reading (V_2 mL)	Volume of H_2SO_4 consumed {($V_1 - V_2$) mL}
Phenolphthalein alkalinity	A1				
	A2				
	A3				
Methyl orange alkalinity	B1				
	B2				
	B3				

Calculation:

(I) Phenolphthalein alkalinity (P) in mg/L as
$$CaCO_3 = (V \times N \times 50 \times 1000)/(\text{Sample volume})$$

(II) Methyl orange alkalinity/total alkalinity (T) in mg/L as
$$CaCO_3 = (V \times N \times 50 \times 1000)/(\text{Sample volume})$$

where V = mL of H_2SO_4 titrant consumed and N = Normality of H_2SO_4

Depending on the result of titration, different types of alkalinity can be calculated from the equation given in the following table:

Titration result	Types of alkalinity (in mg/L as $CaCO_3$)		
	Hydroxide	Carbonate	Bicarbonate
P = 0	0	0	T
P < ½T	0	2P	T – 2P
P = ½T	0	2P	0
P > ½T	2P – T	2(T – P)	0
P = T	T	0	0

Result:

(I) Phenolphthalein alkalinity (P) in mg/L as $CaCO_3$ = ...
(II) Methyl orange alkalinity/total alkalinity (T) in mg/L as $CaCO_3$ = ...
(III) Hydroxide alkalinity (mg/L) = ...
(IV) Carbonate alkalinity (mg/L) = ...
(V) Bicarbonate alkalinity (mg/L) = ...

Environmental significance: High alkalinity of water makes the taste bitter and not suitable for consumption. Chemical treatment often increases the pH value of the water. Alkalinity measurement is necessary for corrosion control; assessment of the effectiveness of coagulants. It gives a measure of the buffer capacity of waste water to neutralize the effect of acid rain. Excess alkalinity in water is a threat to irrigation which reduces crop yield. Alkalinity of water also decides the responsiveness of the waste waters to bio-treatments. An alkalinity value of <250 mg/L is suitable for domestic consumption.

3.8 DISSOLVED OXYGEN (DO) IN WATER

All natural waters and wastewaters contain some dissolved oxygen (DO) in it depending on the physical, chemical and biochemical activities prevailing in the water body. The presence of oxygen (O_2) is essential for the survival of aquatic life. The major sources of DO in natural water are from atmosphere and photosynthetic reactions. Dissolved oxygen is also required in waste water for aerobic degradation and stabilization of organic substances by bacteria and microorganisms.

The solubility of O_2 in water depends on temperature, pressure, chloride concentration, altitude, etc. The solubility of atmospheric O_2 in water varies from 14.6 mg/L at 0°C to about 7 mg/L at 35 °C under one atmospheric pressure. Low DO in water possesses threats to aquatic life. Moreover, a low DO level in a water body is one of the first signs of organic pollution. Thus, measurement of DO is important for assessment of the quality of water and plays significant role in controlling water pollution.

Experiment 3.4

Aim: Determination of dissolved oxygen (DO) in water sample using Winkler's (azide modification) method.

Principle: The chemical determination of dissolved oxygen concentrations in a water sample is done based on the Winkler's (iodometric) method modified by Parsons and Strickland. Oxygen present in the water sample oxidizes iodide (I^-) to iodine (I_2) quantitatively. The amount of I_2 liberated is then determined by titrimetric method using a standard thiosulfate ($S_2O_3^{2-}$) solution as titrant. The end point is denoted with starch indicator. The amount of DO can then be calculated from the titre volume—one equivalent of O_2 reacts with four equivalents of thiosulfate.

In this process, at first, dissolved oxygen is fixed by adding Mn(II) to the water sample under basic conditions, resulting in precipitation of manganic hydroxide [$MnO(OH)_2$, brown colour precipitate]. Then the water sample is acidified to pH 1.0–2.5 which results in dissolution of hydroxide precipitates, liberating Mn^{3+} ions which oxidize the added I^- ions to generate equivalent amount of I_2. This liberated I_2 combines with surplus I^- ions to form I_3^- complex. I_2 and I_3^- complex exist in equilibrium and this complex acts as a pool of I_2. The amount of iodine is then determined by titration with thiosulfate. The stoichiometric equations for the reactions involved in the total process are given below.

$$Mn^{2+} + 2OH^- \rightarrow Mn(OH)_2$$
$$2Mn(OH)_2 + \tfrac{1}{2}O_2 + H_2O \rightarrow 2MnO(OH)_2 \text{ (in alkali medium); fixation of the oxygen}$$
$$MnO(OH)_2 + H^+ \rightarrow Mn(OH)_3 \text{ (in acidic medium, pH = 1.0–2.5)}$$
$$2Mn(OH)_3 + 2I^- + 6H^+ \rightarrow 2Mn^{2+} + I_2 + 6H_2O$$
$$I_2 + I^- \rightarrow I_3^-$$
$$I_3^- + 2S_2O_3^{2-} \rightarrow 3I^- + S_4O_6^{2-}$$

Azide modification: The iodometric method of determination of DO is greatly affected by presence of oxidizing and/or reducing agents present in solution. Oxidizing agents also oxidize I^- to I_2 causing positive interference while reducing agents reduce I_2 back to I⁻ causing negative interference. Bio-treated effluents from treatment plants of wastewater, incubated BOD samples and river water contain nitrite (NO_2^-) ions in considerable amount which also oxidize I⁻ to I_2 in acidic medium leading to erroneous results.

$$2NO_2^- + 2I^- + 4H^+ \rightarrow I_2 + N_2O_2 + 2H_2O \text{ (acidic medium)}$$

This N_2O_2 is again oxidized by DO to NO_2^- establishing a cyclic process.

$$N_2O_2 + \tfrac{1}{2}O_2 + H_2O \rightarrow 2NO_2^- + 2H^+$$

Hence, with interference from nitrites, it is impossible to obtain a permanent end point. This problem can easily be overcome with use of sodium azide (NaN_3), which is incorporated in the alkali-KI reagent. Thus, when the medium is made acidic, following reactions happen:

$$NaN_3 + H^+ \rightarrow HN_3 + Na^+$$
$$HN_3 + NO_2^- + H^+ \rightarrow N_2 + N_2O + H_2O$$

Apparatus: 300 mL BOD bottle, volumetric flask, conical flask, burette, pipette, measuring cylinders.

Reagents:

(i) *Manganese sulfate solution:* Dissolve 480 g $MnSO_4 \cdot 4H_2O$ in distilled water, filter, and dilute to 1 L in a volumetric flask. The $MnSO_4$ solution should not give a colour with starch when added to an acidified potassium iodide (KI) solution.

(ii) Alkali-iodide-azide reagent.

(iii) Conc. H_2SO_4.

(iv) *Starch solution:* Dissolve 2 g laboratory-grade soluble starch and 0.2 g salicyclic acid as preservative in 100 mL hot distilled water.

(v) *Standard sodium thiosulfate titrant (0.025 M):* Dissolve 6.205 g $Na_2S_2O_3 \cdot 5H_2O$ in distilled water and add 0.4 g solid NaOH and dilute to 1000 mL. Standardize with bi-iodate solution.

(vi) *Standard potassium bi-iodate solution (0.0025 M):* Dissolve 974.7 mg $KH(IO_3)$ in distilled water and dilute to 1000 mL.

Collection of sample: Fill the BOD bottle (300 mL) with the given water sample. Then add 1 mL $MgSO_4$ solution followed by the addition of 1 mL alkali-iodide-azide to the BOD bottle. During this addition, the pipette tip should be below the water level. Stopper the BOD bottle with care to exclude air bubbles and make the solution homogenous. Then shake it well allowing sufficient time for all DO to react with the added chemicals. After that, keep the bottle undisturbed for few minutes so that the precipitates settle down leaving above a minimum of 100 mL clear solution. Immediately after precipitation, this mixture is used for measuring the amount of DO in the sample.

Procedure:

(i) *Standardization of thiosulphate:* Dissolve about 2 g KI in a conical flask with 100 mL distilled water. Then add few drops of conc. H_2SO_4 followed by 20 mL (V_1 mL) standard bi-iodate solution. The solution becomes brown in colour. Dilute the solution to 200 mL. The liberated iodine is titrated with the prepared thiosulfate titrant. When the solution becomes pale yellow, add starch indicator, resulting solution will be blue in colour. Continue the titration till the end point when the blue colour of the solution disappears. Record the volume of thiosulfate solution consumed (V_2 mL).

(ii) *Estimation of DO:* Add 2 mL of concentrated H_2SO_4 to the previously prepared sample solution. Re-stopper the bottle and mix the solution well by tender inversion until the precipitate is dissolved completely and a uniform yellow colour solution is obtained. Take out 203 mL of this solution from the BOD bottle in to a conical flask. As 1 mL each of $MgSO_4$ and alkali-KI-azide reagent and 2 mL H_2SO_4 have been added, the proportionate quantity of yellow colour solution in the BOD bottle corresponds to 200 mL of water sample is [= (200 × 300)/(300 – 4) =] 203 mL. Titrate the 203 mL mixture solution with sodium thiosulphate solution until a colour change is observed from dark yellow to pale yellow. Add 1–2 mL starch solution which gives a blue colour to the solution. Continue the titration until the end point is detected by the disappearance of the blue colour. Note down the volume of thiosulphate solution consumed (*V* mL).

Observations:

Table 3.2: Standardisation of thiosulphate

S. No.	Volume of bi-iodate solution (V_1 mL)	Strength of solution (M)	Initial burette reading (mL)	Final burette reading (mL)	Volume of thiosulphate consumed (V_2 mL)

Table 3.3: Determination of DO in water sample

S. No.	Description of sample (mL)	Initial burette reading (mL)	Final burette reading (mL)	Volume of thiosulphate consumed (V mL)

Calculation: Strength of thiosulphate titrant = V_1 × Strength of bi-iodate solution/V_2.

Remember that in 200 mL sample, consumption of 1 mL sodium thiosulfate of 0.025 M is equivalent to 1 mg/L dissolved oxygen concentration.

Hence, dissolved oxygen (DO) (in mg/L) = ... mL of sodium thiosulfate (0.025 M) consumed.

Results: The dissolved oxygen present in the sample = ... mg/L.

Environmental significance: The amount of DO depends upon the temperature of water. Solubility of O_2 in water decreases with increase in temperature. The amount of DO in fresh water varies from 14.6 mg/L at 0 °C to about 7 mg/L at 35 °C under 1 atm pressure. In general, aquatic life requires 2 to 5 mg/L of DO in water. The various kinds of chemical and biological impurities decrease the DO level in water, whereas algae family in water increases the DO level during their photosynthesis process. Drinking water should have a high value of DO to taste good though water with high DO value is corrosive in nature to metals.

The organic matter present in the wastewater requires oxygen for its degradation, called biochemical oxygen demand (BOD) which is satisfied by the DO in water. When BOD is greater than DO, all DO is consumed in the biodegradation process. This depletion in DO level causes death of aquatic living beings which in turn increases the BOD value. Hence, it is very essential to measure the DO of waste water and treat the water properly to lower the BOD before disposal. Moreover, determination of the DO value is a must for aerobic biotreatment of wastewater to control the aeration rate.

BIBLIOGRAPHY

1. Chandra Bhushan and DD Basu. *Gauging the Ganga: Guidelines for sampling and monitoring water quality*, 2017, Centre for Science and Environment, New Delhi.
2. Purity W. Water Purity–Myths and Challenges. *Science in Parliament*, 2013, 70(3), 36.

QUESTIONS

Multiple Choice Questions

1. Which of the following is an aesthetic water contaminant?
   (a) Copper   (b) Dirt
   (c) Pesticide   (d) Cadmium
2. Borosilicate containers may contaminate water by incorporating .... to it.
   (a) Boron   (b) Silicon
   (c) Silica   (d) Borate
3. To eliminate bacterial inhibition during sterilization, which of the following chemical is used?
   (a) NaCl   (b) $Na_2S_2O_3$
   (c) HNO_3 (d) Na_2SO_4
4. Which of the following is not a physical method of water purification?
   (a) Distillation   (b) Filtration
   (c) Sedimentation   (d) Flocculation
5. pH is defined as:
   (a) Logarithm of hydrogen ions
   (b) Negative logarithm of hydrogen ions
   (c) Hydrogen ion concentration   (d) OH ion concentration
6. The product of H^+ and OH^- ions concentration of pure water at 25°C is:
   (a) 10^{-7} (b) 10^{-14}
   (c) 10   (d) 10^7
7. The acceptable value of pH of drinking water is:
   (a) 7.0–8.5   (b) 6.5–9.5
   (c) 6.0–8.5   (d) 6.5–10
8. The favourable pH range for alum treatment is:
   (a) 6.5–9.5   (b) 6.0–9.0
   (c) 6.5–8.5   (d) 7.0–7.5
9. The favourable pH range for bio-treatment is:
   (a) <5.0   (b) >10.0
   (c) 5.0–10.0   (d) ~7.0
10. To determine the mineral acidity and total acidity of water titration should be carried out to pH .... and ...., respectively.
    (a) 7.0, 8.5   (b) 8.5, 7.0
    (c) 8.3, 3.7   (d) 3.7, 8.3

11. The determination of mineral acidity and total acidity of water uses ...... and ...... indicator, respectively.
    (a) Methyl orange, phenolphthalein   (b) Phenolphthalein, methyl orange
    (c) Methyl orange, EBT   (d) EBT, methyl orange
12. Results of acidity analysis are expressed in mg/L of ....
    (a) $MgCO_3$ (b) $CaCO_3$
    (c) Na_2CO_3 (d) H_2CO_3
13. The concentration of dissolved oxygen (DO) in water is mainly dependent on:
    (a) Temperature   (b) Chloride concentration
    (c) Organic purity of water   (d) All of the above
14. How solubility of oxygen in water is related to temperature?
    (a) It decreases with increase in temperature
    (b) It increases with increase in temperature
    (c) It decreases with decrease in temperature
    (d) It does not depend on temperature
15. Generally the required DO for aquatic life is in the range of:
    (a) 12 ppm   (b) 8.5–10 ppm
    (c) 6–8.5 ppm   (d) 2–5 ppm

Answers

1. (b); 2. (c); 3. (b); 4. (d); 5. (b); 6. (b); 7. (a); 8. (c); 9. (c); 10. (d); 11. (a); 12. (b); 13. (a); 14. (a); 15. (d)

Practice Questions

1. What is pure water? Discuss briefly various sources of water contamination.
2. Classify different water contaminants and discuss each type of contaminant briefly.
3. What is water sampling? Briefly discuss the sampling methods.
4. What is water purification? Discuss any two methods briefly.
5. Define pH. Describe the principle of potentiometric method for pH measurement.
6. A decrease in pH of 1 unit represents how much increase in hydrogen ion concentration?
7. What is the importance of pH measurement of water?
8. What is meant by acidity of water? Discuss the source and nature of acidity.
9. Describe the principle of acidity measurement.
10. What pH range is used to (a) measure mineral acidity and (b) total acidity in water?
11. Discuss the significance of acidity measurement.
12. What is meant by alkalinity in water? Describe the principle of its measurement?
13. Discuss the significance of alkalinity measurement. What is the permissible limit of alkalinity in drinking water?
14. What is dissolved oxygen (DO)? Discuss the environmental significance of measurement of DO.
15. Describe the principle of DO measurement.

4

Analysis of Food

UGC Syllabus

Nutritional value of foods, idea about food processing and food preservations and adulteration.
1. Identification of adulterants in some common food items like coffee powder, asafoetida, chilli powder, turmeric powder, coriander powder, pulses, etc.
2. Analysis of preservatives and colouring matter.

INTRODUCTION

The most basic need of any living being is food. It is vital for sustenance and nourishment of life. Food is the source of the basic constituents (carbohydrate, protein, fat) of our body structure and important molecules and ions (vitamins and minerals) for maintaining our immune systems.

The branch of chemistry which deals with the study of analytical procedures to characterize the properties of foods and their constituents is known as food analysis. This study provides information about a safe, nutritious desirable food and also makes the consumers aware of different adulterants (impurities) that may be present in a food item to make it below grade and not suitable for long-time consumption.

4.1 NUTRITIONAL VALUE OF FOODS

The quality of a food item depends on its nutritional value which refers to the impact of the constituents on our body. The major nutrients of food are classified as proteins, fats, carbohydrates (sugars, dietary fibre), vitamins, and minerals. Nutritive value of a food item plays indispensable role in maintaining performance of neonatal, growing, finished and breeding animals. In general, the label on a food container provides information about the nutritional values of that food item to consumers.

• *Nutritional value of carbohydrate:* Each meal of the day should include some carbohydrate rich foods, i.e. the grain group including cereal, rice, or bread. The main energy currency of our body is the carbohydrates. Grains are also rich in vitamins of B group and selenium which play important roles in protecting our body from infection and oxidative damages, proper functioning of the reproductive system. Grains are also good source of dietary fibre which is not digested and contributes very low amount of calories. The main function of such carbohydrates is to make the stool bulky so that it becomes easy to pass. Fibres also help in control the cholesterol level by preventing some cholesterol from being engrossed into the bloodstream.

The most interesting thing to know about the grain group is that when they are processed, only the starchy endosperm is taken and the germs and brans are removed although these parts contain the most of the fibre and nutrients. Alternatively, whole grains are not deprived of nutrients. They are rich in vitamin, minerals and fibre.

• *Nutritional value of protein foods:* Protein is the second most important nutrient which is needed to be included in a meal. Apart from the animal protein such as fish, chicken or egg vegetables like beans, nuts and seeds are also good source of proteins. Proteins and amino acids are the building units of the body, constituting the muscle, organ, bone, skin, and hair tissues. Proteins also play crucial role for healthy functioning of the immune system. Proteins are also rich in vitamins (B and E), iron, magnesium and zinc.

• *Healthy fats and dairy foods:* The fats present in our body are of two types; the unsaturated fats (mono and poly) found in oils like canola and olive oils are good for the health whereas trans and saturated fats present in junk food and fatty meats are bad. Good fats are essential for the proper functioning of our heart and brain. It lowers the bad LDL (low-density lipoprotein) cholesterol level and raises the level of healthy HDL (high-density lipoprotein) cholesterol. Healthy fats are available in seeds, nuts and fatty fish.

Dairy products are rich in calcium, which is essential for strong bones, and vitamin D that works with calcium to build bone health. Fermented dairy products, like yogurt are good sources of probiotics. Milk, ghee, cheese and other dairy products contain high level of saturated fat. Hence, it is advisable to choose non-fat and low-fat versions of dairy products and limit the intake of high-fat dairy products like sour cream, cheese, etc.

• *Fruit and vegetable benefits:* Fruits and vegetables are considered to be the powerhouse of nutrients. They are rich in vitamins, minerals, fibre and antioxidants content which are very essential for our immune system. Thus we should include considerable amount of fruits and vegetables in our meals to improve our health. The nutritious benefits of fruits and vegetables in terms of the constituents are described below:

– Vitamin C: Acts as antioxidant and is essential for collagen synthesis.
– Vitamin A: The vitamin is essential for eyesight, and also supports immune function, reproduction.
– Potassium: Required for nerve and muscle function. Balances the blood pressure.
– Folate: A significant vitamin for growth development and avoidance of birth defects.

People of different age, physiology and gender require different nutrients. Depending on the physical health and activity one should maintain a proper balance diet. Lower and higher dietary intake may result in diseases related to undernutrition and overnutrition, respectively. India uses a pyramid symbol providing guidelines about recommended amount of different kinds of food. The food pyramid is shown in Fig. 4.1. At the base of the pyramid there are cereals and beans/legumes which are suggested to be eaten in plenty. The vegetables and fruits on the second level should be consumed generously whereas animal proteins and oils on the third level should be eaten moderately. At the apex of the pyramid there are highly processed foods which are rich in fat and sugar content should be consumed sparingly.

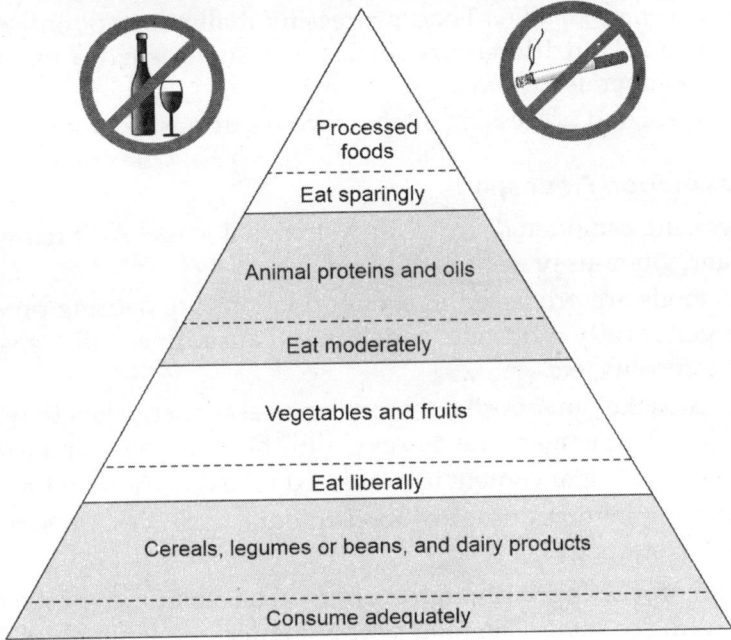

Fig. 4.1: Food pyramid

4.2 FOOD PROCESSING

Food processing is the method of conversion of agricultural substances into food products having particular textural, sensory and nutritional properties. It enhances their shelf life and also makes them functionally more useful. The method of processing foods involves several traditional techniques (heating, smoking, drying, curing, pickling, and fermentation) along with some modern methods (high pressure processing, ultra-heat treatment, and pasteurisation). According to the number of treatments applied, processing methods can be of three types:

Primary processing: The transformation of raw ingredients to food commodities is called primary processing. The basic cleaning, cutting, sorting, milling, grading, and refrigeration are primary processing techniques.

Secondary processing: Secondary processing converts the ingredients obtained from primary processing into ready to use products.

Tertiary processing: Tertiary food processing is the method of large scale commercial production of ready-to-eat foods.

Advantages of Food Processing

- *Increasing availability:* Food processing can safely preserve and package foods, so that it can be transported through the globe and, therefore, it is available everywhere in all seasons.
- *Ensuring food safety:* Consuming good quality of food is very essential for proper maintenance of health. Processing can reduce the incidence of food-borne diseases. Foods are processed to kill the pathogenic microorganism to make them safer and stable for longer time.

- *Preserving nutritional quality:* Food processing maintains not only the quality of food, but also its nutritional properties. It also adds extra nutrients such as vitamins and minerals to foods.
- *Improving taste:* Food processing often improves the taste of foods.

Disadvantages of Food Processing

- Food processing can often decrease its nutritional value. As it removes nutrients like fibre and vitamins present in the food.
- Generally, foods are processed in chemical laboratory. During processing some foods are genetically modified which may cause illness like gastrointestinal disorders, infertility, etc.
- Natural food intake can refresh our mood, increases energy levels, whereas taking processed foods for longer time makes people become angry and irritable.
- The trans-fats and sugar content in processed foods can cause inflammation.
- Food processing involves usage of food colours, additives, preservatives which are harmful to our body.

Methods of food processing: Although every food has their own processing method, the basic principles are same. The food processing technique involves the following general, mechanical and chemical methods.

(I) *Handling and storage of ingredients:* The source of food ingredients includes plant and animal products. The optimum temperature, humidity and personal hygiene should be well maintained during handling the raw materials. The ingredients are then stored in hopper, bins, tanks, cold stores, etc.

(II) *Extraction:* Depending on the ingredient types, extraction procedure is applied. The extraction methods such as crushing, grinding are applied for cereals, liquids and fruits. Direct heating is applied for the extraction of cocoa, coffee and chicory.

(III) *Addition of food additives:* The extracted food ingredients and food additives are accurately weighed and mixed according to the formula. The chemical substances which are used to preserve foods or to improve the quality, flavour, colour, texture, consistency appearance without affecting the characteristics of foods are called *food additives.* After that the food items are treated via processes like cooking, dehydration, fermentation and distillation. For giving the food products definite shapes rolling, cutting, extrusion, etc. are the processes involved.

(IV) *Packaging and transport:* In final step of food processing, the treated foods are packed in suitable container with added preservative. Finally, they are labelled and transported throughout the globe.

4.3 FOOD ADULTERATION

Food adulteration is an act of intentionally degrading the quality of a food item by mixing inferior substances or removing some valuable nutrients from the food.

Adulterants refer to the material use of which could make the food unsafe or substandard (Table 4.1).

Identification of adulterants in some common food items like coffee powder, asafoetida, chilli powder, turmeric powder, coriander powder, pulses, etc.

Table 4.1: Adulterants in some other common food items

Food item	Adulterant
Milk and milk product	
(i) Milk	Water, starch, removal of fat, glucose, sugar, sodium bicarbonate/neutralizer, urea, boric acid, vanaspati, formalin, detergent, sodium chloride
(ii) Ghee/butter	Vanaspati or margarine, mashed potatoes, sweet potatoes and other starches
Oils and sweetening agents	
(i) Mustard oil	Cotton seed oil
(ii) Coconut oil	Any other oil, cyanide
(iii) Sugar	Chalk powder, urea
(iv) Honey	Sugar solution, invert sugar/jaggery
(v) Jaggery	Sodium bicarbonate, metanil yellow colours, washing soda

(I) Coffee Powder

Adulterant: *Chicory*

Method for detection: Sprinkle a pinch of the suspected coffee powder on the surface of water taken in a glass. The coffee floats whereas the chicory sinks down within few seconds leaving behind a trail of colour resulting from the caramel they contain.

Adulterant: *Cereal starch*

Method for detection: Take a pinch of coffee powder in a test tube containing 3 mL of distilled water. Heat the test tube to get a coloured solution. Add ~3 mL of $KMnO_4$ solution followed by HCl (1:1) resulting in decolouration. The formation of blue colour in the mixture, when adding a drop of 1% aqueous solution of iodine indicates adulteration with starch.

Adulterant: *Tamarind seed/date*

Method for detection: Sprinkle the suspected coffee powder on white filter/blotting paper and spray 1% sodium carbonate solution on it. Tamarind and date seed powder will, if present, stain blotting paper/filter paper red.

Adulterant: *Corched persimmon*

Method for detection: Take 1 tsp of coffee powder and spread it on a moistened blotting paper. Pour 3 mL 2% aqueous Na_2CO_3 solution carefully and slowly on it. Appearance of red colour indicates the presence of scorched persimmon powder.

(II) Asafoetida (Hing)

Adulterant: *Soap stone or other earthy mailer*

Method for detection: Take a little amount of hing in water, shake well and allow to settle. Soap stones or any other earthy mailer will settle down.

Adulterant: Starch

Method for detection: Blue colouration on addition of tincture iodine confirms the presence of starch.

Remarks: Compound asafoetida contains starch and it is declared on the manufacturer's level. Hence the test is not applicable for compound asafoetida.

Adulterant: *Foreign resin*

Method for detection: Pure asafoetifda burns like camphor. Take a small amount of sample on a spoon; if it burns like camphor, the sample is pure.

Remarks: Pure hing burns like aromatic camphor.

(III) Chilli Powder

Adulterant: *Brick powder, salt powder or talc powder*

Method for detection: Take 1 tsp chilli powder in a glass of water. If the water becomes coloured, artificial colours are present. If the sediment at the bottom of the glass contains rough particles brick or sand powder is present and if the sediment feels smooth and soapy, soap stone is present in the chilli powder.

Take a pinch of chilli powder and add few drops of conc. HCl. Mix well to make a paste. Dip a stick into the paste hold over the flame. Brick red flame confirms the presence of calcium salts in brick powder.

Adulterant: *Artificial colours*

Method for detection: Sprinkle the chilli powder on a glass of water. Artificial colourants descend as coloured streaks.

Adulterant: *Oil soluble coal tar colour*

Method for detection: Take 2 g sample in a test tube and add few mL of ether. Shake well and separate the ether layer into a test tube and add 2 mL of dil. HCl. Shake well; if oil soluble colour is present as an adulterant, the lower acid layer will be pink in colour.

Adulterant: *Water soluble synthetic colour*

Method for detection: Sprinkle a small quantity of spice powder on the surface of water kept in a glass. Water soluble colours drop down as colour streaks.

Adulterant: *Sudan III*

Method for detection: Weigh 1 g of the chilli powder in a test tube and add 2 mL of hexane. Shake it well. Allow the solution to settle for few minutes and decant the clear solution into another test tube. Add 2 mL of acetonitrile (ACN) and shake the mixture well. The lower layer of ACN layer becomes red colour in presence of Sudan III.

Adulterant: *Sawdust*

Method for detection: Sprinkle chilli powder on a glass of water. The sawdust will float on the water surface.

Adulterant: *Rhodamine B*

Method for detection: Take ¼ tsp of chilli powder in a test tube containing 3 mL water. Then add 10 drops of CCl_4. Shake properly to mix all the contents well. The red colour disappears on shaking. If the red colour reappears on addition of one drop of HCl, the chilli powder is adulterated with rhodamine B.

(IV) Turmeric Powder

Adulterant: *Coloured saw dust (metanil yellow)*

Method for detection: Take 1 tsp of turmeric powder in a test tube and add few drops of conc. HCl. The colour immediately changes to pink. If the colour disappears on dilution with H_2O, the turmeric is pure but if the colour persists, metanil yellow adulteration is present.

Adulterant: *Chalk powder or yellow soap stone powder*

Method for detection: Take a pinch of turmeric powder in a test tube with small quantity of water. If effervescence is observed on addition of a few drops of conc. HCl, turmeric powder is adulterated with chalk or yellow soap stone powder.

Adulterant: *Starch of maize, wheat, topica, rice*

Method for detection: A microscopic study shows that pure turmeric is of yellow colour, big in size and has angular structure, while starches from other sources are small in size and colourless.

(V) Coriander Powder

Adulterant: *Dung powder*

Method for detection: Take 5 g of coriander powder and add it to water. Dung will float and can be easily detected by its foul smell.

Adulterant: *Common salt*

Method for detection: To 5 mL of sample, add a few drops of silver nitrate. White precipitate indicates adulteration.

(VI) Mustard Seed

Adulterant: *Argemone seed*

Method for detection: Mustard seeds have smooth surface whereas the argemone seeds have rough surface and they are black in colour. Hence, they can be easily separated out by close examination. Mustard seeds are yellow colour from inside and argemone seeds are white in inside.

(VII) Black Pepper

Adulterant: *Papaya seeds*

Method for detection: Papaya seeds can be separated out from pepper as they are shrunken, oval in shape and greenish brown or brownish black in colour.

Adulterant: *Coated with mineral oil*

Method for detection: Black pepper coated with mineral oil gives kerosene like smell.

(VIII) Pulses, viz. Dals, like Moong, Channa

Adulterant: *Lead chromate*

Method for detection: Shake 5 g of pulses with 5 mL of water and add few drops of HCl. Pink colouration indicates lead chromate.

(IX) Besan/Yellow Dal

Adulterant: *Kesari dal (Lathyrus sativus)*

Method for detection: Add 50 mL of dil. HCl to 10 g of sample and cook the mixture for about 15 min. If pink colour develops it indicates the presence of Kesari dal.

Adulterant: *Metanil yellow*

Method for detection: Take a pinch of besan in a test tube and add 3 mL alcohol. Shake the test tube to mix the contents properly. Add few drops of HCl to it. Generation of a pink colour solution confirms the presence of metanil yellow.

(X) Wheat Flour

Adulterant: *Excess bran*

Method for detection: Sprinkle on water surface. Bran will float on the surface.

Adulterant: *Chalk powder*

Method for detection: Shake sample with dil. HCl. Effervescence indicates chalk.

Remarks: Chalk powder is used as an adulterant due to its weight.

Adulterant: *Excessive sand and dirt*

Method for detection: Shake a little quantity of sample with about 10 mL of CCl_4 and allow to stand. Dirt and sandy matter will collect at the bottom.

(XI) Green Vegetables like Chilli, Peas, etc.

Adulterant: *Malachite green*

Method for detection: Take a small part of the sample and place it over a moistened white blotting paper. Colour impressions on paper indicate the presence of malachite green.

4.4 FOOD PRESERVATIVES AND COLOURING MATTER

The substances that are not normally consumed as food or component of food with or without nutritive value, which are used for a technological purpose like in the manufacture, processing, preparation, treatment, packing, packaging transport or holding of different food items, are called food additives. The term does not include contaminants or substances added to food for maintaining or improving its nutritive value.

Food additives are purposely added to food and therefore it must be safe for a lifetime of consumption based on current toxicological evaluation. Food additives are used with a purpose of maintaining or improving the keeping quality, texture, consistency, appearance and other technological requirements. The contaminants such as, pesticide residues, metallic contamination, mycotoxins, etc. are excluded from the list. Food additives do not include use of vitamins, minerals, herbs, yeast, hops, starter cultures, malt extract, etc. Food additives are classified on the basis of their functional use. Among different types of food additives we will discuss on preservatives and food colours in this chapter.

Preservatives: Preservatives are the compounds used to prevent and retard the microbial decomposition of foods. Section 3.1.4 of FSS (Food Products Standards and Food Additives) Regulations, 2011 defines preservative as *"a substance which when added to food is capable of inhibiting, retarding or arresting the process of fermentation, acidification or other decomposition of food"*. They are classified into Class I and Class II preservatives (Table 4.2).

Advantages of Food Preservation

Food preservatives are added to keep the food fresh and nutritious by preventing the microbial degradation. Therefore, food preservation has the following advantages:

- It increases shelf life, ensuring round the year availability of seasonal foods.
- The preserved foods can be shipped to far-off distances from the production site.
- It increases the availability of variety of foods.
- It decreases food hazards and spoilage from microbial pathogen.

Table 4.2: Classification of food preservatives	
Class I preservatives	*Class II preservatives*
(i) Common salt	(i) Salts of benzoic acid
(ii) Sugar	(ii) Sulphurous acid and its salts
(iii) Dextrose	(iii) Na and K salts of nitrates and/or nitrites
(iv) Glucose	(iv) Sorbic acid and its Na, K and Ca salts
(v) Spices	(v) Ca or Na propionates
(vi) Vinegar or acetic acid	(vi) Na, K and Ca salts of lactic acid
(vii) Honey	(vii) Nisin
(viii) Edible vegetable oils	(viii) Methyl or propyl parahydroxy benzoates
	(ix) Sodium diacetate

- It prevents the growth of bacteria, fungi, and other microorganisms.
- It retards the oxidation of fats which cause rancidity.
- It inhibits the natural ageing and discoloration of foods.
- It also stabilises food prices since prices tend to rise when supply is low and demand high.
- Preservation process sometime results in variety of products which is very much important for usage in different cuisine. Such as, grapes can be processed in three different ways to give raisins, squash and wines.

Disadvantages of Food Preservation
- The preserved food may cause food poisoning if not stored properly.
- Prolonged use of preserved food can be dangerous to human health.
- It sometime alters the food flavours from its original one.
- Preservation may lead to loss of some nutrients with time.
- It changes the texture and colour of some foods.

Some Traditional Preservation Techniques

(I) *Drying:* The growth of microorganisms like bacteria, yeasts and moulds in food requires presence of water. Hence, reduction of the water content in a food item can solve the problem. Drying is one of the oldest food preservation methods which mainly reduce the water content to retard or prevent the microbial growth in food items. This process not only hinders the quality decay but also reduces the weight of the food. Generally, the water is partially or fully removed by evaporation technique such as sun drying, air or wind drying, smoking and freeze drying. In freeze drying method, food is first frozen and then the water is eliminated by sublimation. There are some other useful drying methods available such as, bed dryers, shelf dryers, spray drying, household oven and commercial food dehydrators.

Many fruits like apples, pears, bananas, mangoes, papaya, apricot, and coconut are dried and preserved for longer time. Raisins which are mainly dried grapes are the most common example of dried foods. For cereal grains such as wheat, rice, maize, barley, oats, and millet, the drying is the usual method of preservation.

(II) *Cooling:* Cooling is also very familiar preservation technique commercially and domestically operated for a variety of food including the prepared food items. The low temperature condition retards the microbial growth on food. Cold stores are used for large scale preservation of the food. Fruits and vegetables are kept in cold stores for long term. For short-term preservation, regular refrigerators are very useful.

(III) *Freezing:* In freezing, the foods are kept in frozen state. Freezing of food slows down decomposition by converting the residual moisture into ice which inhibits the growth of most bacterial species. Meat, fish, some green vegetables like, peas, corns, etc. are preserved by freezing.

(IV) *Heating:* This is an effective method of preserving the food by killing the harmful pathogens at high temperature close to the boiling point of water. Although the food preservation methods such as freezing and heating are comparable, heating is more effective. Milk is boiled before storing to kill microorganisms.

(V) *Sugaring:* The food preservation method using sugar is an old technique. Sugar basically draws the water from the microbes (plasmolysis) and so the microbial cells are dehydrated. Due to dehydration microbial spoilage occurs and the foods are saved from degradation. Using the sugar solution or syrup many fruits like apples, peaches, pears, plums, apricots are preserved. The crystallised sugar is also used as preservatives for citrus fruit (candied peel), angelica and ginger. The sugar is often blended with alcohol for preservation of luxury products such as fruit in brandy or other spirits. Honey is also a common preservative to store fruits. Jam sugar or gelling sugar contains pectin (gelling agent) and citric acid (preservative). It is not only required to make jams and jellies set properly but it acts as preservative also.

(VI) *Pickling:* Pickling or salting is a common preservation method which draws moisture from the food products through osmosis using pickling agent, an edible anti-microbial liquid. The method can be broadly classified into two types such as chemical pickling and fermentation pickling.

In chemical pickling, the pickling agent kills the microorganisms or prevents their growth. Some common pickling agents are brine (high in salt), alcohol, vinegar, and vegetable oil (olive oil). Sometimes, this process also requires heating or boiling so that the food being preserved becomes saturated with pickling agent. Cucumbers, peppers and eggs, mixed vegetables like achar, etc. are the common chemically pickled foods.

In fermentation pickling, the food product itself yields the preserving agent, such as lactic acid. Kimchi, a traditional Korean fermented vegetable food is an example of fermented pickled food.

In commercial pickles, preservative like sodium benzoate or EDTA may also be added to increase shelf life.

(VII) *Lye:* Lye is a strongly alkaline solution such as sodium hydroxide solution which makes the food too alkaline for bacterial growth. Lutefisk is made from air-dried whitefish or dried salted cod pickled in lye.

(VIII) *Canning:* In 1790, French confectioner Nicolas Appert invented this preservation method. Later on, the French Navy successfully preserved variety of foods such as vegetables, fruit, meat, and milk using the same technique. In this method foods are kept in a sterilized jar or containers and heated to a temperature that kills the microorganisms which cause food spoilage. Finally, the hot containers are sealed and cooled. The vacuum sealing process helps to kill any remaining bacteria within the jar or can.

Different foods have different degrees of natural protection against degradation. High-acid fruits such as strawberries need a short boiling cycle without preservatives, whereas, low-acid foods, such as vegetables and meats need pressure canning. The preserved food using canning technique can have immediate risk of spoilage once the container has been opened.

Some Modern Industrial Techniques

(I) *Pasteurization:* This is a unique preservation technique which is not intended to remove all the pathogenic microorganisms from the food item to be preserved. Pasteurization process reduces the number of viable pathogens so that they do not cause any disease (assuming pasteurized food items are stored properly and consumed before its expiration date). Commercial-scale sterilisation process affects the quality and taste of the food. Hence, the technique is less practiced. Flash pasteurization which is comparatively a newer preserving method involves short time exposure of the food items to high temperatures, and is claimed to be better for preserving quality and taste in some eggs.

(II) *Vacuum packing:* In this packaging method, air is removed from the packet to have the vacuum environment inside the bag or bottle for storing foods. The main point is to remove oxygen with intimate contact between the packaging material and food item. This type of preservation method is commonly used for storing nuts, cereals, chesses, chips, etc. to prevent oxidation which causes the loss of flavour and quality.

(III) *Irradiation:* Irradiation method involves the exposure of food to ionizing radiation such as X-rays from accelerators, high-energy electrons, gamma rays (emitted from radioactive sources as Co^{60} or Cs^{137}), etc. The treatment has a wide range of effects, including killing bacteria, moulds and insect pests, reducing the ripening and spoiling of fruits, and at higher doses inducing sterility. The technique is comparable to pasteurization and often called 'cold pasteurization'. Irradiation is not effective to protect the food from viruses and toxins already formed by microorganisms. It is only useful for food with high initial quality.

Some other preservation techniques are jugging, jellying, pulsed electric field processing, irradiation, high pressure, modified atmosphere, bio-preservation and burial in the ground.

Benzoic Acid

Experiment 4.1

Aim: Qualitative and quantitative estimation of benzoic acid in food items.

(I) *Qualitative detection*: Acidify the food sample with HCl acid (1:3) and extract the organic contents in diethyl ether. Remove the solvent from the ether extract completely on a hot water bath. Dissolve the residue in few mL of hot water and add few drops of 0.5% ferric chloride solution. Formation of salmon colour precipitate of ferric benzoate indicates the presence of benzoic acid in the sample food.

(II) *Quantitative estimation*:

Titrimetric method:

Principle: Benzoic acid is separated from a known quantity of food sample. A solution is made with the given amount of food sample and the solution is saturated with sodium chloride followed by acidification with dil. HCl. Benzoic acid is then extracted with chloroform. The chloroform layer is made mineral acid free and the solvent is removed by evaporation. The residue is dissolved in neutral alcohol and the amount of benzoic acid is determined by titration against standard alkali.

Chemical reagents: Distilled chloroform, dil. hydrochloric acid, 10% sodium hydroxide solution, 0.05 N standard sodium hydroxide solution, saturated sodium chloride solution.

Preparation of sample:

(I) *Beverages and liquid products:* Shake the sample thoroughly to make it homogenous and transfer 100 g of the sample into a 250 mL volumetric flask using saturated solution of sodium chloride. Make the solution alkaline by adding 10% NaOH solution and check with litmus paper. Fill the volume with saturated sodium chloride solution. Mix rigorously and allow the volumetric flask to stand for 2 h. Filter the solution mixture and use the filtrate for determination of benzoic acid.

(II) *Sauces and ketchups:* Weigh 150 g of sample and add 15 g sodium chloride salt. Transfer the mixture into volumetric flask and rinse with saturated NaCl solution. Then add 15 g pulverized NaCl followed by addition of 10 mL of 10% sodium hydroxide solution. Make the volume to 500 mL with sodium chloride solution. Let it stand for 2 h with occasional shaking. Filter the solution and use the filtrate for determination.

(III) *Jams, jellies, preservatives and marmalades:* Mix 150 g of sample with 300 mL saturated NaCl solution. Then add 15 g pulverised sodium chloride followed by addition of 10 mL of 10% NaOH solution. Transfer the mixture to 500 mL volumetric flask and dilute to volume with saturated NaCl solution. Let it stand for 2 h with frequent shaking, filter and use the filtrate for determination.

Procedure: Pipette out 100 mL of the filtrate into a 250 mL separating funnel. Neutralize the solution with hydrochloric acid and add 5 mL excess. Extract the organic substances with portions of chloroform. Avoid emulsion formation by gentle shaking with rotatory motion. If emulsion forms, break it by stirring the chloroform solution with a glass rod after each extraction. Do not drain any of the emulsion with chloroform layer. Transfer the combined chloroform extract into a separating funnel and wash it with water by shaking gently to make the organic layer free from mineral acid. Drain off the water layer. Dry the chloroform layer over anhydrous sodium sulphate and remove the solvent by evaporation. Remove the last traces of the solvent under a current of air at room temperature. Dry the residue overnight or until no residue of acetic acid is detected if the product is a ketchup. Dissolve the residue in 30–50 mL of alcohol (neutralised to phenolphthalein) and titrate with 0.05 N sodium hydroxide solutions.

Calculation of the amount of benzoic acid:

Weight of the sample (W) = ... g

Volume made up (V_1) = ... mL

Volume taken for extracting with chloroform (V) = ... mL

Titre value (V_2) = ... mL

Normality of the NaOH (S) = ... N

Now, 1 mL 1 N NaOH ≡ 122 mg benzoic acid

Therefore, benzoic acid (ppm or mg per kg) $= \dfrac{122 \times V_1 \times V_2 \times S \times 1000}{W \times V}$

Sorbic Acid

Sorbic acid is used as preservative (antimicrobial agent) in dairy products, cottage cheese, ricotta and mozzarella cheese, sour cream, yoghurt, etc. The presence of sorbic acid in food items can be detected qualitatively as well as quantitatively by a single method.

Experiment 4.2

Aim: Qualitative and quantitative estimation of sorbic acid in food items.

Principle: Sorbic acid is extracted from the food sample with a solvent mixture of diethyl ether and petroleum ether (1:1) and absorbance of the extract is measured within 200–300 nm range. Sorbic acid gives a characteristic peak at ~250 nm. If the extract of the food sample is treated with permanganate and optical spectrum is recorded, absence of the peak at 250 nm is taken as confirmation of the presence of sorbic acid in the sample.

Again for quantitative estimation, a standard curve is drawn by recording the UV spectra of pure sorbic acid solutions with varying concentration at 250 nm. Then the electronic spectrum of the food extract is recorded. Comparing the value of absorbance of the food extract at 250 nm with the standard curve, we can easily quantify the amount of sorbic acid present in the food sample.

Apparatus: Volumetric flask, measuring cylinder, pipette, separating funnel, filter paper (Whatman No. 1 and Whatman No. 3), quartz cuvette, spectrophotometer.

Chemical reagents:

(i) *Metaphosphoric acid solution:* Dissolve 5 g in 250 mL water and dilute to 1 L with alcohol.

(ii) *Mixed ether solvent:* Petroleum ether and anhydrous diethyl ether (1:1).

(iii) *Potassium permanganate ($KMnO_4$) solution:* Dissolve 15 g in 100 mL water.

(iv) *Sorbic acid standard stock solution (1 mg/mL):* Dissolve 100 mg of sorbic acid in mixed ether and make up to 100 mL volume in a volumetric flask.

(v) *Working standard solution of sorbic acid:* Dilute 5 mL of the previous solution to 100 mL with the mixed ether solvent.

(vi) *Reference solvent for optical spectra:* Shake 10 mL of mixed ethers with 100 mL of phosphoric acid solution and dry the supernatant ether layer with anhydrous sodium sulphate.

Sample preparation: Homogenise the sample (cheese and related products) by cutting into small pieces or blending with a high speed blender (for creamed cottage and similar cheeses). Accurately weigh about 10 g of the homogenised food sample, in a high speed blender, add enough phosphoric acid to yield a total of 100 mL of liquid in the mixture. Blend for one minute and immediately filter through Whatman No. 3 paper. Transfer 10 mL of filtrate to a 250 mL separating funnel and extract the organic substances with 100 mL of mixed ether solvent. Discard the aqueous layer and dry the ether extract over 5 g of anhydrous sodium sulphate.

A. *Qualitative detection of sorbic acid procedure*:

Procedure: Take 10 mL of the ether extract and add 2 mL $KMnO_4$ solution. Shake for one minute. Filter the organic layer through Whatman No. 1 paper and dry the filtrate over anhydrous sodium sulphate. Record absorbance spectra of the solution between 220 nm and 300 nm. Absence of peak at 250 nm confirms the presence of sorbic acid.

B. *Quantitative estimation of sorbic acid by spectrophotometric method:*

Procedure: Prepare four standard solutions of sorbic acid by adding 1, 2, 4 and 6 mL of the standard stock solution to four different 100 mL volumetric flasks and dilute to volume with mixed ethers. Label the solutions as P, Q, R and S. Determine the absorbance of each solution (P, Q, R and S) at 250 nm against mixed ether solvent.

Draw a standard curve by plotting absorbance (A) values against concentration of sorbic acid (mg per 100 mL). Now, determine the absorbance of the sample solution and calculate, the concentration of sorbic acid in the sample solution from the standard curve. Express the final result in ppm.

Calculation: The percent amount of sorbic acid present per gram of food sample can be calculated from the following equation:

$$\text{Sorbic acid (\%)} = \frac{\text{Amount of sorbic acid (mg)}}{\text{Amount of sample (mg)} \times 1000} \times 100$$

Result: The amount of sorbic acid present as preservative in the given food sample is = ... ppm

• *Food colours:* To identify the synthetic food colouring agents, first they have to be separated from food items. The colouring matters are of two types—(a) natural and (b) synthetic colours. They may be further classified as water soluble and oil soluble. Natural colours are isolated or synthesized from natural sources. Different types of natural food colours and their identification procedures are listed in Table 4.3.

Table 4.3: Different natural food colours and their detection methods

Colour	Source	Detection method
Caramel	Carbohydrates (corn, wheat and sugar cane)	*Fiehe's reaction:* Extract the caramel from the food caramel with 50 mL ether and pour on a porcelain dish and evaporate to dryness. Add 3 drops of 1% solution of resorcinol in HCl to the residue. The presence of caramel is indicated by appearance of rose red colour.
Cochineal	Scale insect	Dissolve the colour material in amyl alcohol and shake it with dilute ammonia. Generation of a purple colour solution confirms the presence of cochineal.
Curcumin	Turmeric	Evaporate the alcoholic extract of the material to dryness on the water bath with a piece of filter paper. Moisten the dried filter paper with few drops of weak boric acid solution and add few drops of HCl acid and dry the paper again. The presence of curcumin is confirmed if the dry paper shows cherry red colour, which changes to bluish green by a drop of NaOH or NH_4OH.
Annatto	Seeds of the achiote tree	Shake the melted fat or oil with 2% NaOH solution and pour the aqueous extract on moistened filter paper. The filter paper will show a straw colour which will persist with a gentle wash with water. Dry the paper, add one drop of 40% $SnCl_2$ solution and dry it again carefully. If the colour turns purple, the presence of annatto is confirmed.
Chlorophyll	Plant leaf	Extract the sample with ether and treat the ether extract with 10% methanolic solution of KOH. Colour becomes brown, quickly returning to green, confirms the presence of chlorophyll.
Betanin	Beet root	Extract the aqueous suspension with amyl alcohol. It remains in aqueous phase. Dye it with a piece of tannin mordanted cotton, a terracotta shade is produced in presence of betanin.

Synthetic colours are important as they are widely used in various foods. Eight coal-tar food colours are permitted to be used in certain food products under the provisions of FSS (Food Products Standards and Food additives) Regulations, 2011. The details of the synthetic colours are listed in Table 4.4.

Table 4.4: Different synthetic food colours permitted by FSS

Colour shade	Common name	Colour index	Chemical class
Red	Ponceau 4R	16255	Azo
	Carmoisine	14720	Azo
	Fast red	16045	Azo
Yellow	Tartrazine	19140	Pyrazolone
	Sunset yellow FCF	15985	Azo
Blue	Indigo carmine	73015	Indigoid
	Brilliant blue FCF	42090	Triphenylmethane
Green	Fast green	44090	Triphenylmethane
	Green FCFs	42053	Triphenylmethane

However, there are some unpermitted colours such as metanil yellow, rhodamine B, orange G, blue VRS, auramine, etc. which are often present in food items as adulterants.

BIBLIOGRAPHY

1. Manual of methods of analysis of foods, *Food Additives*, 2015, Food safety and standards authority of India (FSSAI).
2. Manual of methods of analysis of foods, *Instruction Manual Part-II*, 2012, Food safety and standards authority of India (FSSAI).

QUESTIONS

Multiple Choice Questions

1. The substance essential for proper growth is called a:
   (a) Nutrient
   (b) Carbohydrate
   (c) Calorie
   (d) Fatty acid

2. Carbohydrates, proteins and fats present in food as:
   (a) Micronutrients
   (b) Macronutrients
   (c) Both a and b
   (d) None of these

3. Vitamins and minerals are needed by our body as:
   (a) They give energy to the body
   (b) They help carry out metabolic reactions
   (c) They insulate the organs of body
   (d) They draw the heat from the body

4. Which one is not a nutrient?
   (a) Vitamins
   (b) Minerals
   (c) Fibres
   (d) Fats

5. The rice, bread or cereal is good source of:
   (a) Vitamins
   (b) Minerals
   (c) Carbohydrates
   (d) Protein

6. Vegetables and fruits are considered as good source of:
   (a) Vitamins
   (b) Minerals
   (c) Fibre and antioxidants
   (d) All of the above

7. Which of the following is an advantage of food processing?
   (a) Availability of seasonal food throughout the year
   (b) Removal of toxins and preserving food for longer
   (c) Adds extra nutrients to some food items
   (d) All of the above mentioned

8. Which of the following is a method of food processing?
   (a) Handling and storage
   (b) Extraction
   (c) Addition of food additives
   (d) All of the above

9. Which of the following is a disadvantage of food processing?
   (a) Canning of food leads to loss of vitamin C
   (c) Processed food adds empty calories to food constituting junk
   (b) Some chemicals make the human and animal cells grow rapidly which is unhealthy
   (d) All of the above mentioned

10. Pulses, beverages and turmeric are commonly adulterated with:
    (a) Soap and stone
    (b) Papaya seeds
    (c) Metanil yellow
    (d) All of these

11. Coffee powder is mainly adulterated with:
    (a) Chicory
    (b) Metanil yellow
    (c) Papaya seed
    (d) Sawdust

12. Among the following which is not a class-I preservative?
    (a) Sugar
    (b) Dextrose
    (c) Acetic acid
    (d) Benzoate salt

13. Canning of fruits and vegetables is a .... process.
    (a) Cold
    (b) Heat
    (c) Irradiation
    (d) Microwave

14. The preservation technique used to remove microorganisms from food is:
    (a) Canning
    (b) Irradiation
    (c) Pasteurization
    (d) Vacuum packing

15. Beverages, pickles, jams, cheeses, salads contain .... as preservative.
    (a) Benzoate
    (b) Sorbates
    (c) Both a and b
    (d) None of these

16. Which one is not a natural food colour?
    (a) Indico carmine
    (b) Cochineal
    (c) Annatto
    (d) Caramel

Answers

1. (a); 2. (b); 3. (b); 4. (c); 5. (c); 6. (d); 7. (d); 8. (d); 9. (d); 10. (c); 11. (a); 12. (d); 13. (b); 14. (c); 15. (c); 16. (a)

Practice Questions

1. What do you mean by nutritional value of foods?
2. What is food processing? Write down the advantages and disadvantages of it.
3. Briefly discuss the different methods of food processing.
4. Define food adulteration. What adulterants are used in coffee powder, asafoetida and chilli powder?
5. How do you identify metanil yellow adulteration in turmeric powder?
6. How do you identify malachite green in green vegetables like chilli, peas, etc.?
7. Define food preservatives. Give two examples of each class I and class II preservative.
8. What are the different traditional techniques for food preservation? Discuss any two of them.
9. Write a short note on: (i) Pasteurization, (ii) Canning, (iii) Sugaring, and (iv) Vacuum packing
10. Write down the advantages and disadvantages of food preservation.
11. Write down the principle of qualitative and quantitative determination of benzoic acid and sorbic acid in food items.
12. Classify different food colours with example.
13. Write down the source and detection method of the following natural colours: (i) caramel and (ii) curcumin, and (iii) betanin

5

Chromatography-I

INTRODUCTION

Among various physical and chemical methods to separate two or more chemical compounds from a mixture, chromatography is the technique best fitted for compounds which closely resemble each other in their physical and chemical properties (such as mixtures of amino acids). Chromatography is superior to other separation techniques because of the use of a wide variety of procedures, materials, and equipment. Due to these advantages chromatography is the most widely used separation technique in almost all chemical laboratories.

5.1 BACKGROUND OF CHROMATOGRAPHY

Chromatography is a non-destructive method for separating different constituents (major, minor and even trace) of a multi-component mixture. The technique was first employed in 1906 by a Russian botanist **Dr Mikhail Tswett**. He used the technique for

Fig. 5.1: 2D and 3D chromatographic techniques

separating chlorophyll, xanthophyll and several other colour pigments present in a vegetable extract using a $CaCO_3$ column. The different extent of adsorption of the pigments on the adsorbent ($CaCO_3$) plays the key role in the separation technique. In this experiment **Tswett** obtained coloured bands at different positions of the adsorbent ($CaCO_3$) column. He named this band of colours as the 'chromato-gram' and the technique as 'chromatography' after the Greek words 'chroma' and 'graphs' meaning 'colour' and 'writing', respectively.

The Nobel Prize winners in Chemistry, 1952

Archer John Porter Martin   Richard Laurence Millington Synge

The popularity of chromatography technique in modern scientific age can be attributed to **Martin** and **Synge** for introducing partition column chromatography in 1941. In later years, **Martin** and co-workers also developed two different separation techniques—paper chromatography and gas chromatography. The Nobel Prize in Chemistry 1952 was awarded jointly to **Archer John Porter Martin** and **Richard Laurence Millington Synge** "for their invention of partition chromatography".

Definition: According to **Dr Mikhail Tswett**, *chromatography is a process used for separating substances by filtering their solution mixture through a column of adsorbent and then developing the column with a solvent.*

Chromatography can also be defined as *the technique for separation of components from a mixture of solutes by using dynamic partition of dissolved materials between two immiscible phases having different polarity.*

Chromatography is defined by IUPAC as a *"physical method of separation in which the components to be separated are distributed between two phases, one of which is stationary (stationary phase) while the other (the mobile phase) moves in a definite direction"* (MacNaught and Wilkinson, 1997).

5.2 CLASSIFICATION OF CHROMATOGRAPHY

Based on the nature of mobile phase chromatography methods are mainly classified in two categories—gas chromatography and liquid chromatography. These two techniques again use different kind of stationary phases and various separation mechanisms depending upon which numerous chromatographic methods are developed. A tabular format of this classification is provided in Flowchart 5.1. In this book, we will mainly focus on chromatographic methods which involve liquid mobile phase and solid stationary phase. Thin layer chromatography (TLC), paper chromato-graphy (PC), column chromatography (CC) and ion-exchange chromatography (IEC) methods are included in the newly proposed CBCS syllabus by UGC, India as these techniques are widely used and practiced in almost all analytical laboratories. This chapter will focus on liquid–solid chromatographic methods using planar (2D) stationary phase such as thin layer chromatography (TLC) and paper chromatography (PC). Detailed principles and applications of column chromatography (CC) and ion-exchange chromatography (IEC) using column (3D) as stationary phase will be discussed in Chapter 6.

Flowchart 5.1: Classification of chromatography technique

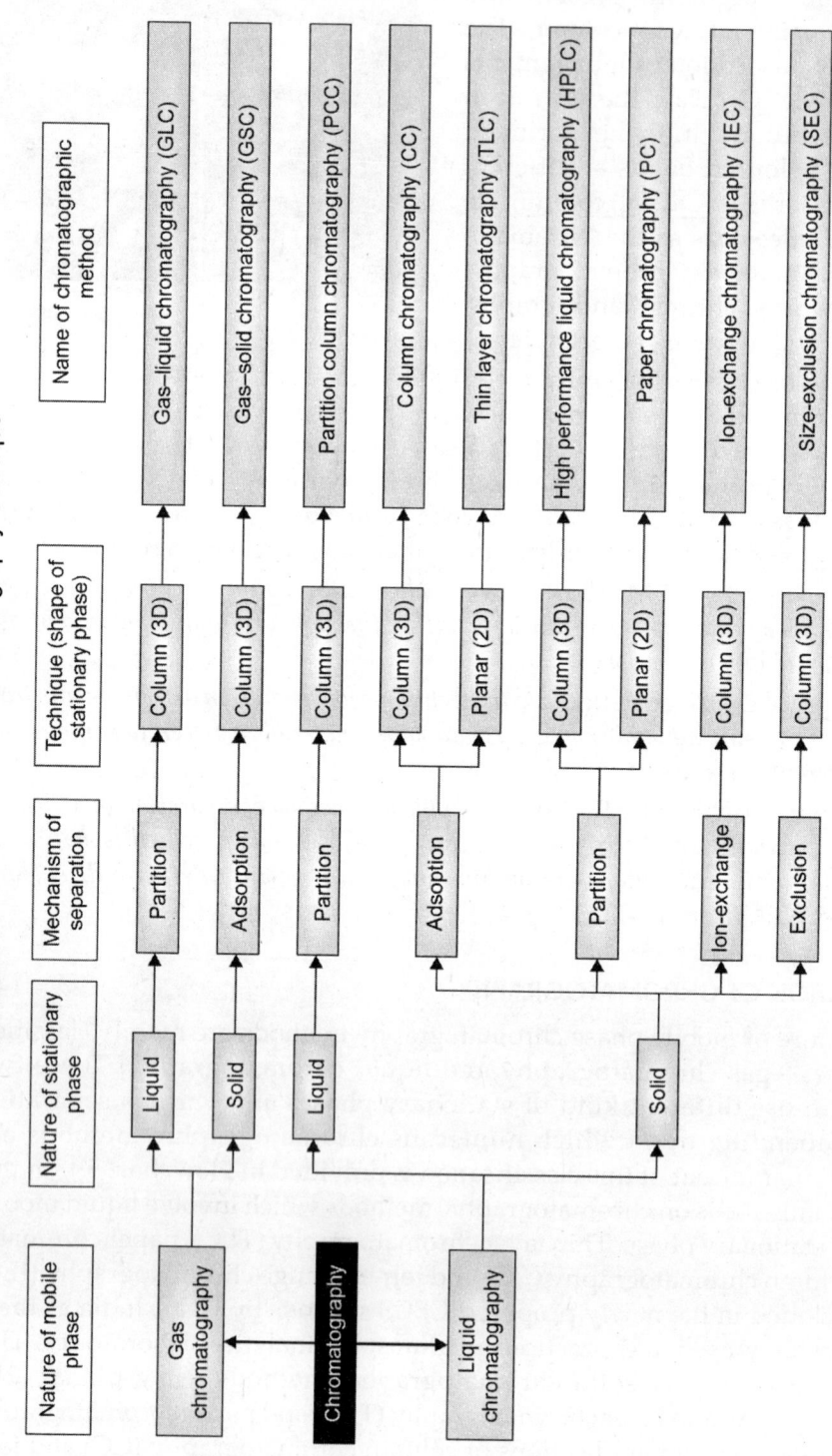

5.3 PAPER CHROMATOGRAPHY

Paper chromatography is an analytical tool to separate coloured chemicals or substances. This two-dimensional (2D) chromatography method was first developed by Martin and Synge in 1944. In paper chromatographic technique paper sheets or strips are used as the stationary phase (adsorbent) and a solvent or solvent mixture is generally used as mobile phase (eluent) (Fig. 5.2). This method is helpful in separating chemical substances by using the differential migration rates of the compounds on the sheet of the paper due to the difference in their partition coefficient values. This technique is very much helpful in analytical chemistry that uses very small quantities of material.

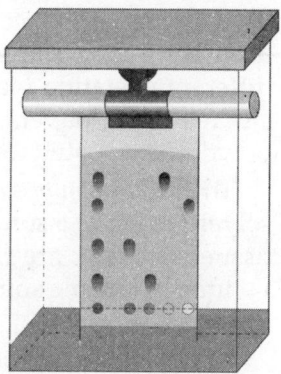

Fig. 5.2: Paper chromatography technique

Principle: Paper chromatography is one kind of partition chromatography where the substances are partitioned or distributed between two phases. The two phases are solvent held in pores of the filter paper (stationary phase) and the solvent which develops through the paper (mobile phase). In this method a drop of the mixture solution is casted on the paper strip (generally filter paper) and dried to create a small spot. The paper is then kept in a closed chamber with the lower edge dipped into the mobile phase. As the filter paper gets the liquid through its capillary axis, the mobile phase moves upward. The components of the mixture migrate with the developing solvent at different rates creating spots at different point on the paper. After the mobile phase travelled to a suitable length on the paper strip, the paper is dried. The spot corresponding to coloured compounds are easily located on the chromatographic paper. However, for detection of spots of colourless compounds, it may be treated with suitable chemical reagents called visualizing agent or developer which gives the spot a characteristics colour. Iodine is the most commonly used developer in paper chromatography. The individual components are identified by calculating the retardation factor (R_f) and comparing with that of the standard substances. The R_f values can be calculated by using the following expression:

$$R_f = \frac{\text{Distance travelled by the substance from reference line (cm}}{\text{Distance travelled by the solvent front from reference line (c}}$$

General procedure: The experimental procedure of paper chromatography technique involves several steps. These steps are discussed in detail below.

(I) *Choice of stationary phase*: In paper chromatography, a filter paper generally Whatman filter paper containing ~99% of α-cellulose is used as stationary phase. A suitable chromatographic paper should have good clarity and separation efficiency, minimum diffuseness of the spot, and good mobility of the mobile phase through it.

Depending on the nature of chromatographic work, the chemical composition of the filter paper is modified. For separating polar compounds the exchange efficiency of the paper is improved by increasing the carboxyl content (1.4%) by partial oxidation. Again partial hydrolysis of the filter paper on treatment with 7% HCl solution for 24 h followed by washing with DI water and ethanol provides increased capillarity to the stationary phase. Several adsorbents like alumina, zirconium oxide, silica, etc. are often used as coatings on cellulose paper to produce a thin flexible sheet of adsorbent to be used as stationary phase in paper chromatography. Treatment of the filter paper with different chelating reagents such as dimethylglyoxime (DMG), 8-hydroxyquinoline, etc. results in papers with special features. The filter papers are also infused with powdered or liquid ion-exchangers to fabricate ion-exchange papers.

(II) *Preparation of sample:* A small quantity of solid sample is dissolved in suitable solvent to make a solution which can directly be applied on the paper. For biological tissues, the cells are treated with suitable solvent to prepare an extract. A 10–20 μL volume of sample solution containing microgram amount of substance is enough for spotting.

(III) *Application of sample (spotting):* A piece of filter paper of 15–30 cm in length and one to several centimetres in width is selected depending on the number of samples to be spotted. A pencil line is drawn ~2 cm apart from one edge, this line is called origin or baseline. On the baseline some marks are made 2 cm from each other. A drop of each sample is transferred with the help of capillary tubes on the mark of the baseline to create spots of about 0.5 cm in diameters. Minimum quantity of solution should be used to avoid diffusion. The spots are dried by allowing the solvent to evaporate or a blow dryer is used for quick drying.

(IV) *Choice of mobile phase:* The selection of the solvent for mobile phase depends on the nature of the substances to be separated. However, the solvents should have following characteristics:

- Solvents should not be toxic or should not have carcinogenic effect.
- For solvent mixtures, the composition should remain unaltered for longer period of time.
- Solvents should not have any chemical reactivity with any of the components of the sample.
- Solvents should not produce any interference in detection of the spots on the developed paper.
- The solvent for mobile phase is so chosen that the differences in R_f values of any two components should be at least 0.05 for differentiating two closely spaced spots. The R_f values of the sample should be in the range of 0.05 to 0.85.

Some commonly used solvent mixtures in paper chromatography are—butanol : acetic acid : water (4 : 2 : 12); acetic acid : methanol (3 : 1); ethanol : isopropanol : HCl (9 : 9 : 2); ethyl acetate: n-butanol: acetic acid: water (80 : 10 : 5 : 5).

(V) *Development of chromatogram:* The chromatogram can be developed either by ascending or by descending technique (Fig. 5.3). The ascending technique allows the solvent to travel up the paper while in descending technique the solvent runs down the paper. Depending on the nature

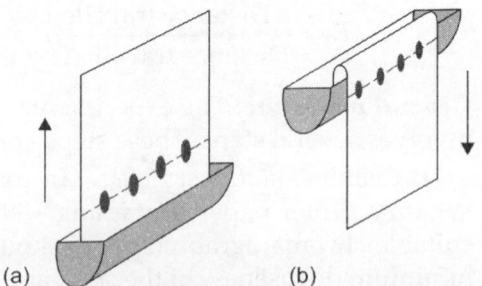

Fig. 5.3: Development techniques in paper chromatography: Ascending (a) and descending (b)

of the substances to be separated suitable development technique is chosen.

After spotting the filter paper is immersed in the developing chamber according to the selected development technique. The solvent begins to move and pulls the components of the sample differentially along with it. At the end of the development the chromatographic paper is taken out and the solvent front is marked with another line. The chromatographic filter paper is dried by blowing hot air or any other suitable method.

(VI) *Detection of compounds:* Colour compounds can be easily detected on the paper but for colourless compounds various chemical and physical methods are adopted for clear detection.

• *Physical methods:* The physical methods for detection of compounds in paper chromatography include fluorescence or radioactivity assay. Some compounds invisible on chromatogram under ordinary light can easily be detected under UV lamp because of their fluorescence property. Most of the unsaturated organic compounds produce characteristic emitted light which helps in identification of the compound. Though a wide range of compounds is stable to UV light, vitamins and steroids are destroyed when UV visualization technique is employed for identification.

The radioactivity on a chromatogram can be measured by Geiger-Müller counter and can be used as a tool for isotope detection in paper chromatography.

• *Chemical methods:* The colourless components can be detected by converting them to coloured compounds on treatment with suitable reagents. The reagents can be solid, liquid or even gas (e.g. H_2S gas is used for detection of metal ions which form coloured sulfide compounds). Liquid reagents can be used directly, whereas solids are dissolved in suitable solvents (such as water, methanol, ethanol, *n*-butyl alcohol or acetone) before application. The reagents can either be spread over the chromatogram with a spraying bottle or the chromatogram itself is dipped into the solution without touching.

Applications of Paper Chromatography

The paper chromatography technique has wide range of applications. Some of them are listed below.

- Paper chromatography is specially used for separating the mixtures with both polar and non-polar constituents. Hence, it is used in analysis of reaction mixtures in laboratories.
- This chromatography technique is also useful in separation of peptides, amino acids, sugar, alkaloids, lipids, etc. in biological samples.
- It is used to assess the amino acid composition of proteins.
- This technique is used to study the process of fermentation and fruit ripening.
- Paper chromatography is used to detect adulterants or contaminants present in drinks and foods.
- This technique is useful to check the control of purity of pharmaceuticals and cosmetics.
- It is used to detect dopes and drugs in humans and animals.
- The paper chromatography has also found application in forensic laboratories to analyse different samples collected from the crime scene.
- The technique is also useful in studying the DNA and RNA sequencing.

Advantages of Paper Chromatography

- Paper chromatography is a simple, cheap and rapid separation technique.
- It requires very small quantity of materials.

- This chromatography technique is equally applicable to both inorganic and organic compounds.
- Unlike other analytical methods paper chromatography does not necessitate much space or equipment.
- It has excellent resolving power.

Limitations of Paper Chromatography

- Paper chromatography is not suitable for large quantity of sample.
- The efficiency of paper chromatography is less for quantitative analysis.
- Complex mixture cannot be separated by paper chromatography.
- Corrosive samples cannot be analysed by paper chromatography as the cellulose gets destroyed.
- This technique is less accurate in comparison with HPLC or HPTLC.

5.4 PAPER CHROMATOGRAPHIC SEPARATION OF MIXTURE OF METAL IONS (Fe^{3+} and Al^{3+})

Experiment 5.1

Aim: Separation and identification of Fe^{3+} and Al^{3+} ions present in a given mixture using ascending paper chromatography.

Principle: To identify the component ions of a sample mixture by paper chromatography the sample is spotted on the filter paper and developed using suitable solvent. Post-development spraying with 1% alcoholic solution of alizarin, different colours are obtained for different metal ions. A reddish pink colour will indicate the presence of Al^{3+} spot and a purple colour indicates Fe^{3+} spot.

Apparatus: Whatman filter paper No. 1, capillary tubes, 1 ruler, pencil, chromatography jar with lid.

Chemical reagents:

    Spotting solution: (i) 1% aqueous solution of FeCl$_3$ with few drops of dil. HCl, (ii) 1% aqueous solution of AlCl$_3$ with few drops of dil. HCl.

    Developing solvent: (i) Butanol : acetic acid : water (4 : 2 : 12) or (ii) acetic acid: methanol (3 : 1) or, (iii) ethanol : isopropanol : HCl (9 : 9 : 2)

    Spraying agent: 1% alcoholic solution of alizarin.

Procedure: Prepare 25 mL of developing solvent mixture (mobile phase) and pour this into a chromatography jar covered with a lid. Wait for few minutes, so that the environment inside the jar becomes saturated with solvent vapour which can provide a better chromatographic separation. Take a strip of chromatography paper. While handling the paper, always wear gloves and hold the side edges as foreign chemical substances (if any) from hand can damage the paper surface and reduce its efficiency. About 2 cm apart from the long edge of the paper draw a pencil line which will indicate the origin. Take three capillary tubes for spotting and use different capillary for each solution. Transfer a drop of each solution on the origin line to mark a spot of about 0.5 cm in diameter and each spot should be about 2 cm apart. Allow the paper to dry and mark corresponding spot with second drop of each solution. Mark each spot on the paper with a pencil. Here, the spotting solutions are (i) 1% aqueous solution of FeCl$_3$, (ii) 1% aqueous solution of AlCl$_3$, and (iii) test sample solution. Carefully place the paper into the chromatography jar, so that paper remains vertically straight. Wait for the developing solvent to rise up the paper up to within 1 cm from the top of the paper. Remove the paper and dry it with air drier or oven. Then hold the paper over the top of an evaporating dish containing conc. NH_3 solution. While the paper is moist with NH_3, spray the paper with 1% alcoholic solution of alizarin. Finally warm the paper strip in oven. A reddish pink colour spot will indicate the presence of Al^{3+} while purple colour spot confirms the presence of Fe^{3+}. Measure the distances travelled by each spot and calculate their R_f values.

Observation and Calculation:

Stationary phase = ...

Mobile phase (developing agent) = ...

Spray reagent (visualizing agent) = ...

Distance travelled by the solvent = ...

R_f (retention factor) = Distance travelled by ion/distance travelled by solvent.

Substance	Distance travelled by the solute (cm)	Distance travelled by the solvent (cm)	R_f value	Color after spraying
Fe^{3+} ion (i)				
Al^{3+} ion (ii)				
Component A in a mixture				
Component B in a mixture				

Results:

The R_f value of Fe^{3+} ion is ....

The R_f value of Al^{3+} ion is ....

The R_f value of component A of experimental solution spot is ...., which is almost equal to the R_f value of Fe^{3+} ion. This confirms that component A in the experimental solution contains Fe^{3+} ion.

Again, the R_f value of component B of experimental solution spot is ...., which is almost equal to the R_f value of Al^{3+} ion. This confirms that component B in the experimental solution contains Al^{3+} ion.

5.5 THIN LAYER CHROMATOGRAPHY (TLC)

Thin layer chromatography (TLC) is an extremely useful, fast, easy-to-use technique to visualize components of a mixture. In 1938, two Russian scientists, **NA Izmailov** and **MS Schreiber** investigated the plant extracts using a thin layer of slurried adsorption medium for the first time. Then in 1941, **MO'L Crowe** reported the same results using a thin layer of adsorbent in a petri dish. After that **JE Meinhard** and **NF Hall** improved the technique by introducing binders to the sorbents in 1949. **Dr Justus G Kirchner** founded the present-day system of thin layer chromatography, by developing the TLC plates with sorbent layers in 1951.

Principle: The thin layer chromatography is performed on a TLC plate. A TLC plate is a sheet of aluminium foil, plastic, or glass, which supports the stationary phase, i.e. a thin layer of a solid adsorbent (usually silica, SiO_2 or alumina, Al_2O_3). The mobile phase is a solvent or mixture of solvents. After spotting the sample mixture on TLC plate, it is kept in a closed jar or beaker (TLC chamber) containing the eluent or mobile

phase. The solvent rises up the TLC plate by
capillary action, dragging the sample mixture
along with it. The components of the sample
are attracted by the polar sites of the adsorbent
surface by electrostatic force and this 'binding'
is reversible. The solvent (mobile phase) also
interacts with the components as well as the
adsorbent. Depending on the polarity and
relative interaction of the three systems, solute,
solvent and adsorbent, different components of
the mixture move with different rate. When the
solvent almost reaches 90% length of the TLC
plate, it is taken out of the TLC chamber and the
solvent front is drawn with a pencil. Then the
TLC plate is allowed to dry by evaporating the
solvent and the components of the mixture are
marked under the UV light. The individual
components are identified by calculating the

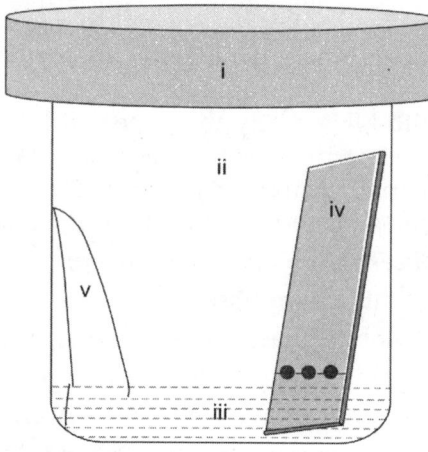

Fig. 5.4: Experimental set up for TLC:
(i) Petri dish, (ii) beaker, (iii) developing
solvent, (iv) TLC plate, and (v) filter
paper

retardation factor (R_f) and comparing with that of the standard substances. The R_f
values can be calculated by using the following expression:

$$R_f = \frac{\text{Distance travelled by the substance from reference line (cm)}}{\text{Distance travelled by the solvent front from reference line (cm)}}$$

General procedure:

(I) *Prepare TLC plates:* A wide variety of TLC adsorbents are commercially available.
Amongst them a TLC plate having aluminum support with silica gel 60 matrix is
commonly used because of its outstanding wettability (even for 100% aqueous
visualization reagents), easy and precise cutting (no flaking of silica) and excellent
separation efficiency. Commercially available large TLC sheets are cut into small size
of 5 cm length and varying width depending upon the number of samples to be
spotted. The plates should be handled carefully. The coating should not be disturbed
or the adsorbent should not be dirty.

Using pencil a line is drawn from ~0.5 cm from the bottom end and this line is origin
line. Some dots are marked (equal to the number of samples to be spotted) on the origin
leaving enough space between two spots. While drawing origin line and marking dots
one should be careful not to disturb or mutilate the adsorbent.

(II) *Spotting of sample:* For solid samples a small quantity is dissolved in suitable
solvent to make a solution which can directly be used for spotting on TLC plate. Using
a capillary tube or a micropipette, a drop of each sample to be analysed is transferred
to the marked dots to create spots of about 0.5 cm in diameters. Minimum quantity of
solution should be used to avoid diffusion. The spots are dried by allowing the solvent
to evaporate or a blow dryer is used for quick drying. For analysis of reaction mixture,
the starting materials should be spotted along with the reaction mixture.

(III) *Choice of mobile phase:* The compounds in the spot will move up on the TLC plate
depending on the solvent chosen for mobile phase. Polar solvents will drag polar
substances while non-polar solvent will make non-polar compounds move with them.
In non-polar solvents like hexane and pentane, most polar compounds will not move,
while in polar solvents non-polar molecules remain bonded with the adsorbent. A

good solvent system is one that transports all components of the analyte mixture from the origin but does not put anything on the solvent front resulting in R_f values in the range of 0.15 to 0.85. This is not always likely, but should be kept at target while running a TLC. Some standard solvent systems used as mobile phase in TLC experiment are given below.

Very polar solvents: Methanol > ethanol > isopropanol

Moderately polar solvents: Acetonitrile > ethyl acetate > chloroform > dichloromethane > diethyl ether > toluene.

Non-polar solvents: Cyclohexane, petroleum ether, hexane, pentane.

Commonly used solvent mixtures: Ethyl acetate : hexane : 0–30% (most popular combination), ether : pentane : 0–40% (very popular), ethanol : hexane/pentane (5–30%), dichloromethane : hexane/pentane (5–30%), chlorobenzene : toluene : 1,2-dichloroethane (1 : 1 : 1 v/v), 1,2-dichlorobenzene : toluene : 1,2-dichloroethane (2 : 1 : 1 v/v).

(IV) *Filling the TLC chamber:* One or two milliliters of the desired solvent or solvent mixture is poured into the TLC chamber. A large piece of filter paper soaked with the solvent is also placed in the chamber. This helps in saturating the atmosphere in the TLC chamber with solvent vapour and stops solvent from evaporating as it rises up the plate.

(V) *Run the TLC:* The TLC plate is carefully placed into the TLC chamber in slightly tilted position so that the origin line remains above the solvent layer. The lid is closed and the solvent is allowed to travel upward to 90% of the way on the plate. Then the TLC plate is removed and the solvent front is drawn with a pencil. The plate is dried by evaporating the solvent.

(VI) *Visualization of spots on TLC plate:* The best method for non-destructive detection of the spots in TLC plate uses a UV lamp. The TLC plate is placed under the UV lamp and the UV active spots are encircled with a pencil. Alternative semi-destructive method involves staining with iodine by placing the TLC chamber in an iodine chamber. The compounds will produce yellow spots on reaction with iodine vapour. After marking the spots in either method the R_f values are calculated.

(VII) A repeat TLC run may be required if R_f value is not good or streaking appears. The solvent polarity is increased if higher R_f value is required or a less polar solvent is chosen if lower R_f value is required. If 'streaking' of compounds occurs on TLC plate, i.e. instead of generating sharp spots, large streaks appear, most probably the sample is too concentrated. The TLC is performed again with diluted sample, if 'streaking' persists one should shift to other choices of solvent.

Applications of TLC: TLC technique is widely used all over the world due to its manifold applications. The various applications of TLC are discussed below.

- *Examination of reaction mixtures:* TLC technique is mostly used to analyze the reaction mixture to examine the advancement of the reaction. Assessment of the reaction mixture by TLC can tell us whether the reaction is complete or not. It can also provide information about separating the different components of the reaction products.

- *In chemistry:* TLC technique is applied for separation, isolation and characterization of organic as well as inorganic compounds. This characterization method is also useful in separating vitamins.

- *Testing sample purity:* TLC can also provide information about the purity of a sample. For this, the test sample and a standard sample are spotted simultaneously and a TLC is run. On development, any impurity present in the test sample will give an extra spot on the chromatograph.
- *Identification of the components of natural products:* Thin layer chromatography can identify and separate different components of the natural products such as essential oil or volatile oil, glycosides, waxes, alkaloids, fixed oil, terpenes, steroids, etc.
- *In pharmaceutical chemistry:* TLC technique is used for qualitative analysis and purity checking of medicines such as analgesics, anesthetics, hypnotics, anti-convulsants, sedatives, tranquillisers, etc.
- *In food and cosmetic industry:* TLC technique is useful in detection of adulterant in foods and impurities in cosmetic products.

Advantages of TLC Technique

- TLC is an extremely useful, easy-to-use, faster and cost-effective separation technique.
- It is highly sensitive and produces more accurate results using least number of equipment.
- TLC involves very small amount of sample for analysis with lower detection limit almost one decimal lower than that of paper chromatography.
- Corrosive compounds can also be analyzed through TLC.
- Diffusion of sample spots is less in TLC.
- The components of samples separated through TLC can be recovered easily by scratching the powdery coating of adsorbent at the spotted area and re-dissolving it. Moreover, quantitative separation of compounds is also possible with TLC.

Disadvantages of TLC Technique

- Disadvantages of breaking the TLC plates include safety hazards and the possibility of producing jagged edges.
- The results of TLC are not easily reproducible.
- TLC is applied for qualitative analysis. It is not so suitable for quantitative appli-cation.
- TLC takes into account only the soluble components of the mixtures.
- The humidity and temperature can affect the results as TLC method is performed in an open system.

The key differences between paper and thin layer chromatography are listed in Table 5.1.

5.6 ANALYSIS OF PAINT SAMPLES BY TLC

Experiment 5.2

Aim: Investigation and separation of the organic pigments in paint sample by TLC.

Principle: To investigate an unknown paint sample, scientists use a comprehensive analysis with known paints by deducing the pigment composition. Thin layer chromatography (TLC) is an important tool for identification and separation of organic

Table 5.1: Key differences between paper and thin layer chromatography

Paper chromatography	Thin layer chromatography
i. The paper chromatography is based on 'partition' principle	i. The thin layer chromatography is based on 'adsorption' principle
ii. The water trapped in the cellulose filter paper serves as the stationary phase	ii. Thin coating of adsorbent (silica or alumina) on glass or metal foil support acts as the stationary phase
iii. Descending techniques are preferred due to lack of physical strength in paper	iii. Ascending techniques are preferred as TLC plates have enough physical strength
iv. Paper chromatography requires less time for particle separation	iv. Thin layer chromatography requires more time
v. Separation is less sharp due to more diffusion of the spots	v. Separation is sharper due to less diffusion of the spots
vi. Non-destructive visualization method using UV lamp is not applicable in paper chromatography	vi. Non-destructive visualization method using UV lamp is used in thin layer chromatography
vii. Corrosive samples cannot be analysed in this method as the corrosive agents can destroy the paper	vii. Corrosive samples can be analysed in this method

pigments in paints. A pigment plays an important role in determining colour and appearance of a paint. To find the discrimination in property, the organic pigments are extracted from the unknown paint first and then a comparison study is carried out using TLC method.

Apparatus: TLC plates, capillary tubes, one ruler, pencil, TLC chamber with lid.

Chemical reagents:

Extracting solution: Dichloromethane or 1,2-dichlorobenzene.

Developing solvent (eluent): Chlorobenzene : toluene : 1,2-dichloroethane (1 : 1 : 1 v/v) or, 1,2-dichlorobenzene : toluene : 1,2-dichloroethane (2 : 1 : 1 v/v).

Paint samples: To prepare paint film clean one glass microscope slide with acetone and dip it into well-mixed liquid paint. After taking out, allow it to dry to get the desired paint film.

Procedure:

Extraction of organic pigments from paint samples: To extract the organic pigments from the paint, take a small portion of the as-prepared paint film having size ~(0.2 × 0.2) cm^2 in a white ceramic cavity tile and add 1 mL of CH_2Cl_2 and keep it for some time. If this extracting solvent is not suitable for this, then decant the solvent and add 1 mL of 1,2-dichlorobenzene to the partly extracted paint fragment. Then keep the tile on a hot plate (temperature should be lower than the boiling point of the solvent) for 15 min.

TLC procedure: Prepare 25 mL of developing solvent mixture (eluent) and pour a little amount of this mixture into a chromatography jar covered with a lid. Wait for few minutes, so that the atmosphere inside the jar becomes saturated with solvent vapour which can provide a better chromatographic separation. Take a TLC plate. While handling the plate, always wear gloves and hold the side edges to avoid any contamination. About 1 cm apart from the bottom edge of the paper, draw a pencil line which will indicate the origin. Take capillary tubes for spotting and use different capillary for each solution. Transfer a drop of each pigment extraction and the known pigment solution on the origin line to mark a spot and each spot should be about 3 cm apart. Allow the paper to dry and mark corresponding spot with second drop of each solution. Carefully place the paper quickly into the chromatography jar, so that paper remains vertically straight and the bottom edge will be immersed in the solvent pool. Wait for the developing solvent to move up the paper up to within 1 cm from the top of the paper. Remove the paper and mark quickly the solvent front (boundary between the dry and wet silica) with the pencil. Measure the distances travelled by each spot and calculate their R_f values.

Calculation: Calculate the R_f values of all the pigment extraction using the following relation.

R_f (retention factor) = Distance travelled by ion/Distance travelled by solvent

Identification of paint pigments: Identify the paint pigments separated by TLC experiment by comparing the obtained R_f values with those of known pigments.

BIBLIOGRAPHY

1. Gulati S , Sharma JL, Manocha S. *Practical Inorganic Chemistry*, 2017, CBS Publishers & Distributors Pvt. Ltd.
2. Home JM, Laing DK, Dudleym RJ. The discrimination of modern household paints using thin layer chromatography. *Journal of the Forensic Science Society*, 1982, 22, 147.

QUESTIONS

Multiple Choice Questions

1. Martin and Synge share Noble prize in Chemistry for their contribution towards development of partition chromatographic technique in:
   (a) 1952   (b) 1953
   (c) 1954   (d) 1955
2. Chromatography separates the mixture of dyes on the basis of their:
   (a) Density   (b) Solubility
   (c) Gravity   (d) Boiling point
3. Paper chromatography is based on the principle of:
   (a) Adsorption   (b) Absorption
   (c) Partition   (d) None of the above
4. The mobile phase in chromatography can be:
   (a) Gas only   (b) Liquid only
   (c) Solid   (d) Gas and liquid
5. In paper chromatography, the most soluble solute:
   (a) does not move from the start point
   (b) stays closest to the start point
   (c) travels furthest away from the start point
   (d) can be found in the middle of the chromatogram
6. Generally, a chromatographic paper is made up of:
   (a) Amylase   (b) Cellulose
   (c) Polymer   (d) None of the above

7. Which is not correct about paper chromatography?
   (a) Used for analysis of mixture of sugar, vitamin, amino acids, etc.
   (b) Used for the identification of drug
   (c) Used for the analysis of metabolites of drug in blood, etc.
   (d) Used in coating

8. Factor(s) that affect R_f value is/are:
   (a) Spot size   (b) Temperature
   (c) Nature of mobile phase   (d) All of the above

9. R_f ranges from:
   (a) 0 to 1   (b) Less than 0
   (c) 1–10   (d) None of the above

10. In ascending paper chromatography, mobile phase moves:
    (a) Towards gravity   (b) Against gravity
    (c) Both a and b   (d) None of the above

11. Ninhydrin solution is used for spot determination of:
    (a) Amino acid and proteins   (b) Fat
    (c) Carbohydrate   (d) All of the above

12. Ammonical silver nitrate solution is used for spot determination of:
    (a) Amino acid   (b) Proteins
    (c) Fat   (d) Carbohydrate

13. Thin layer chromatography is:
    (a) Partition chromatography   (b) Electrical mobility of ionic species
    (c) Adsorption chromatography   (d) None of the above

14. In paper and TLC chromatography stationary phase is:
    (a) Solid phase and liquid phase   (b) Solid phase and solid phase
    (c) Liquid phase and liquid phase   (d) Liquid phase and solid phase

15. The working principle of TLC is based on the concept:
    (a) Different components are absorbed on solid phase to different extent
    (b) Different components are absorbed on solid phase to different extent
    (c) Different components are adsorbed on solid phase to different extent
    (d) Different components are absorbed on solid phase to same extent

16. The spots on the TLC plate are identified by all of the following, except:
    (a) Fluorescence   (b) Spraying with reagents
    (c) Fluorescent adsorbent   (d) Under microscope

17. Which separation technique is useful to identify the pigments in paint sample?
    (a) TLC   (b) Filtration
    (c) Evaporation   (d) Distillation

18. A compound that travels higher on TLC (thin layer chromatography) always:
    (a) Is more polar   (b) Is darker in colour
    (c) Is more conjugated   (d) Has a larger R_f

Answers

1. (a); 2. (b); 3. (c); 4. (d); 5. (c); 6. (b); 7. (d); 8. (d); 9. (a); 10. (b); 11. (a); 12. (d); 13. (c);
14. (d); 15. (c); 16. (d); 17. (a); 18. (d)

Practice Questions

1. What is chromatography?
2. Define stationary and mobile phase.
3. Briefly discuss the principle of paper chromatography.
4. What is meant by the term developing in chromatography? Discuss different development techniques used in paper chromatography?
5. Write a short note on R_f factor.
6. Write the principle of separation of Fe^{3+} and Al^{3+} using paper chromatography?
7. Discuss briefly various applications of TLC.
8. What are the major disadvantages of TLC?
9. What is the purpose of adding filter paper to the TLC chamber?
10. What is the major difference between thin layer chromatography and paper chromatography?

6

Chromatography-II

UGC Syllabus
1. Column, ion-exchange chromatography, etc.
2. Determination of ion-exchange capacity of anion/cation exchange resin (using batch procedure if use of column is not feasible).

INTRODUCTION

Column chromatography technique is the most widely used and user-friendly separation method which enables the separation, identification, and purification of the components of a mixture for qualitative and quantitative analysis. In 1897, DT Day used this technique for the first time to study the different components of natural petroleum. Russian botanist **Mikhail Tswett** (1906) employed the technique for analysis of an organic solution of plant pigments using a vertical glass column packed with an adsorptive material. He observed that on chromatographic separation of the pigment mixture a series of coloured bands appears on the column, separated from each other by colour free regions. Soon, the technique became popular method for separating complex homogeneous chemical mixtures into their individual components.

6.1 COLUMN CHROMATOGRAPHY

According to IUPAC nomenclature, column chromatography is a separation technique in which the stationary bed is within a tube. The particles of the solid stationary phase or the support coated with a liquid stationary phase may fill the whole inside volume of the tube (packed column) or be concentrated on or along the inside tube wall leaving an open, unrestricted path for the mobile phase in the middle part of the tube (open-tubular column). Column chromatography involves adsorption, partition, or ion-exchange phenomenon. This method is suitable for the physical separation of gram-scale compounds.

Principle: Conventional column chromatography technique involves stationary phase consisting of finely divided solid packed in a vertically mounted glass column. The mobile phase generally consists of a single solvent or a solvent mixture which occupies the pores between the solid particles of stationary phase. The sample to be separated usually introduced as slurry either dry or with the mobile phase. The sample is introduced from the top of the column. When the mobile phase moves through the stationary phase the sample components also move with it. Depending on the adsorption interaction of the compounds with the polar sites of the stationary phase

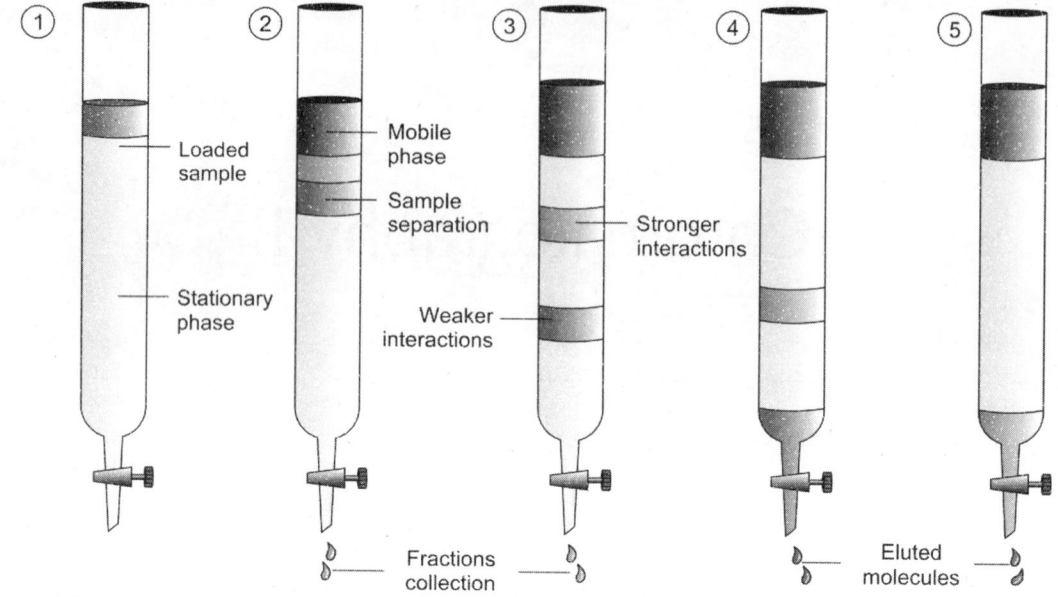

Fig. 6.1: Column chromatography diagram

and difference in solubility of the components in mobile phase ('like dissolves like', i.e. polar compounds will be more soluble in polar solvents whereas non-polar molecules will be dissolved in non-polar solvents), individual components of the mixture moves at different rates. The binding of the compounds with the stationary phase is of reversible type. The compounds with lower adsorption interaction with stationary phase travel faster and will be eluted first. The compounds with greater adsorption affinity with the stationary phase will be eluted later.

The rate of the movement of the components is expressed as:

R_f = The distance travelled by solute/the distance travelled by solvent

R_f is the retardation factor.

General procedure:

(I) *Column:* The adsorption column chromatography generally uses a glass column of diameter from 1 cm to 10 cm and height 20 to 50 cm. The size of the column is chosen according to the amount of the sample to be separated. The smaller the column diameter, the more effective will be the separation. The most commonly used column is thin pyrex walled glass tube of ~15–20 cm length and 1 cm diameter. One end of the tube is drawn out and stoppered. A cotton plug is pressed on the stopper. The plug helps to prevent the adsorbent from falling down but allows the solvent to pass.

(II) *Packing of column:* The column is held vertical using a clamp and the powdered solid adsorbents are packed within the column either by dry packing or wet packing method.

(a) *Dry packing:* Dry powder of adsorbent is introduced into the column through the upper open end. Vacuum is applied from the lower end for well settlement of the adsorbent. While packing it should be ensured that the top is solid and unbroken.

(b) *Wet packing:* Thick slurry of the adsorbent in suitable medium is poured through the open end and allowed to settle by gravitational force until a desired height is obtained.

(III) *Adsorbent:* A large number of choices are available for adsorbent in column chromatography. The adsorbing power of a substance varies significantly according to the fineness, preparation method, activation process, etc. A suitable adsorbent should meet the following criteria.

- Particle should have uniform shape and size in the range of 60–200 μm in diameter.
- High mechanical stability and chemical inertness. The material should be chemically inert to the components of the sample to be analysed and any solvents used in the separation process.
- The compound itself should be colourless, inexpensive and readily available.
- It should allow free passage of mobile phase through it.
- It should be suitable for wide range of samples.

(IV) *Mobile phase:* This phase refers to the solvents used and it accomplishes the following purposes:

- The samples are often introduced into the column with the help of this solvent.
- Solvent separates the different components to form bands in the column. Hence, the solvent acts as a developing agent.
- It also does the job of eluting agent and removes the components of different band from the column.

While choosing a solvent or solvent mixture for mobile phase the polarity and viscosity of the individual solvent should be taken into consideration. For better separation, it is found that the solution mixture should be prepared with relatively non-polar solvent, development should be performed with somewhat polar solvent and even more polar solvent is used to elute the adsorbed materials. Some commonly used solvents are listed below with increasing polarity order:

Petroleum ether < CCl_4 < cyclohexane < CS_2 < ether < acetone < benzene < ester of organic acids < $CHCl_3$ < alcohols < H_2O < pyridine < organic acids.

(V) *Separation of compounds:* Column chromatography involves a very simple technique of separation. A solution or slurry of the sample mixture is made using a non-polar solvent and introduced into the column from the top. If the sample is a coloured compound, it forms a band of uniform height on the top of the adsorbent. The solvent level should remain above the sample layer. The stopcock at the bottom is then opened slightly to allow the solvent to flow. Always the solvent is run until a small amount of solvent remains covering the top of the packed material. After the sample layer settles well, more polar solvent (or solvent mixture) is introduced for development. As the sample solution moves with the solvent down the column different components of the sample moves with different rates. The component which is more tightly adsorbed will move slowly and weakly adsorbed component will move faster. Thus, after some times the compounds form discrete bands in the column. If the compounds are coloured, these bands can be seen through naked eye.

The compounds are then eluted out of the column using even more polar solvent. The compounds of lower bands, i.e. the weakly adsorbed compounds will be eluted first. Elutes of different bands are collected in different conical flasks.

Different components are adsorbed to different extent; they require different time to travel the length of the column. This is known as retention time. Due to difference in retention time, different compounds come out of the column at different times and get separated. The retention time of a compound depends on the following factors:

(i) Nature of the compound, adsorbent, solvent and their relative interaction.

(ii) Column size (diameter and length)

(iii) Mode of packing of column

(iv) Temperature

(VI) *Identification of compounds:* Coloured compounds are easily identified. For colourless compounds the identification process takes the help of UV-light for UV active conjugated organic molecules or appropriate stain for UV-inactive compounds.

In general large number of fractions of eluate of equal volumes is collected in test tubes and each fraction is identified using thin layer chromatography. Fractions that do not contain any components are rejected. Fractions containing the desired compound are identified and all of the fractions that contain the desired isolated compound(s) are merged into a round-bottom (RB) flask. The solvent is evaporated on a rotary evaporator to get the crude component. This procedure is performed for each components of the mixture.

Factors affecting column efficiency: The efficiency of a column in chromatographic separation depends on the following factors:

- *Dimension of the column:* The efficiency of a column depends on its size. Column efficiency increases with increasing length/width ratio.
- *Particle size of adsorbent:* With decreasing size of the adsorbent particles column efficiency increases.
- *Quality of the adsorbent:* The quality and purity of the adsorbent molecules greatly affect the column efficiency. Poor quality adsorbents result in lower column efficiency.
- *Packing of column:* Packing of the adsorbent particles has also an effect on column efficiency. Packing can be done either by dry or wet methods. But in any case, the packing should be uniform and there should not be any air channel within the packed column.
- *Temperature:* The elution speed increases at higher temperature.
- *Solvent:* Quality and purity of solvent also influence the column efficiency. Since flow rate is inversely related to the viscosity, low viscous solvents are used to obtain high efficiency separations in column chromatography.

Applications of column chromatography: Column chromatography is the most useful separation and purification technique for solid and liquid sample mixtures. Its major applications are:

- Separation and isolation of active ingredients of natural products.
- It is a useful technique for impurity removal from a particular compound.
- It is used to determine formulations of a drug.
- This technique is useful in isolation of metabolites in biological fluids.
- It can successfully separate diastereomers.

Classification of column chromatography:

- **Based on flow rate of mobile phase:** In column chromatography, generally a glass tube is used in which fine powder of solid adsorbent is packed. The mobile phase moves through this column of adsorbent with varying rate depending upon the method of elution. Based on this flow rate, column chromatography is classified into five categories:

(I) *Gravity column chromatography:* In this chromatographic method, the mobile phase flows only by the influence of gravitational force (Fig. 6.2a).

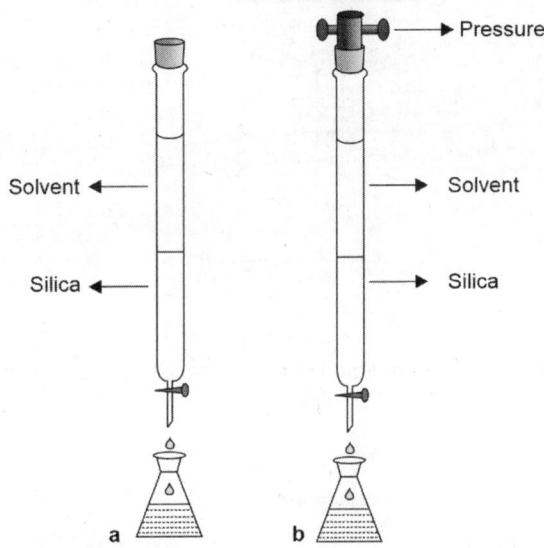

Fig. 6.2: Gravity (left) and flash (right) column chromatography

(II) *Flash column chromatography:* This method uses special valve to push the mobile phase downwards by stream of air or nitrogen (Fig 6.2b).

(III) *Low and medium pressure column chromatography:* The technique uses a pump fitted at the upper end of the column to create pressure so that the mobile phase moves faster. Increased flow rate reduces the time for separation.

(IV) *Vacuum liquid chromatography (VLC):* The rate of flow of mobile phase is increased by creating vacuum. The dry adsorbent is packed in a sintered glass funnel. The sample is introduced either as dry solid or as a solution. Mobile phase is then added on the top by portion and after each addition vacuum is applied to collect each portion (Fig. 6.3).

(V) *High performance/pressure liquid chromatography (HPLC):* In this method very fine powder of silica is used as adsorbent while the flow rate of mobile phase is severely decreased. This method uses a stainless steel tube as column. High pressure pumps are used to retard the velocity of the mobile phase through the silica column (Fig. 6.4).

• **Based on mechanism:**

(I) *Adsorption column chromatography:* This column chromatographic technique involves adsorption of the components of the sample mixture on the polar sites of the adsorbent surface.

(II) *Partition column chromatography:* Both the stationary phase and the mobile phase are liquids and the mechanism of separation involves partition of the components of the sample among these two phases depending on the partition coefficient values of the components in those two solvent systems.

(III) *Gel column chromatography:* In this method of column chromatography, the

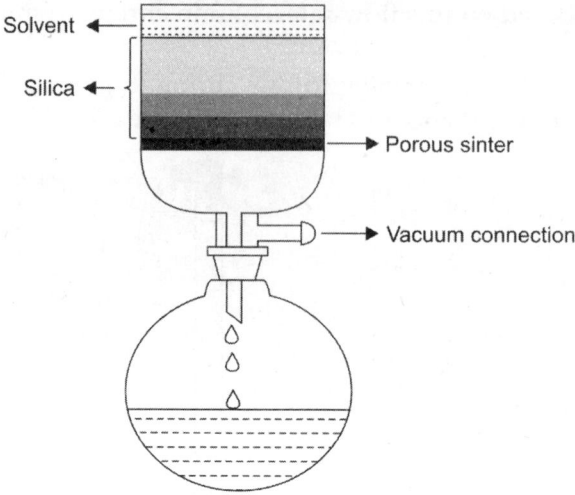

Fig. 6.3: Vacuum liquid chromatography (VLC) diagram

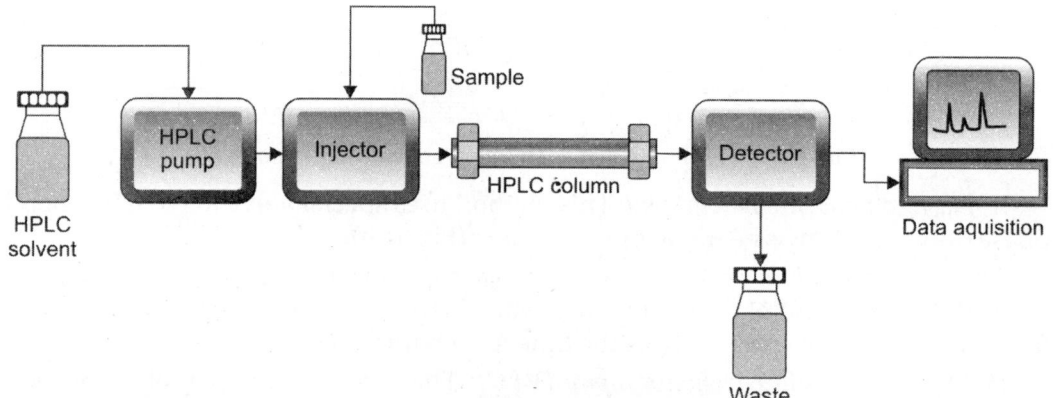

Fig. 6.4: High performance liquid chromatography diagram

separation is performed using a packed column of gel. The stationary phase is the solvent held in the pores of the gel.

(IV) *Ion-exchange column chromatography:* This column chromatography technique uses ion-exchange resins as stationary phase.

S. No.	Column chromatography technique	Mobile phase	Stationary phase	Sample phase
1	Adsorption column chromatography	Liquid	Fine powder of solid adsorbent	Solution
2	Partition column chromatography	Liquid	Immiscible solvent on solid matrix	Solution
3	Gel column chromatography	Liquid	Solvent present in the pores of a gel polymer	Solution
4	Ion-exchange column chromatography	Liquid	Ion-exchange resin	Solution

6.2 ION-EXCHANGE CHROMATOGRAPHY

Ion-exchange chromatography is a high resolution chromatography technique useful in separating complex mixtures (usually proteins and enzymes). It is a special case of column chromatography utilizing ion-exchange resins for column packing. In this chromatography technique the reversible interaction between charged molecules in the mobile phase and ion-exchange matrix with opposite charges results in the binding and elution of specific molecules to accomplish the separation effect. Exchangeable matrix counter ions may include H^+, OH^-, singly charged cations (Na^+, K^+, Cl^-), doubly charged cations (Ca^{2+}, Mg^{2+}), and polyatomic inorganic ions (PO_4^{3-}, SO_4^{2-}) as well as organic acids (COO^-) and bases (NR_2H^+) (Fig. 6.5).

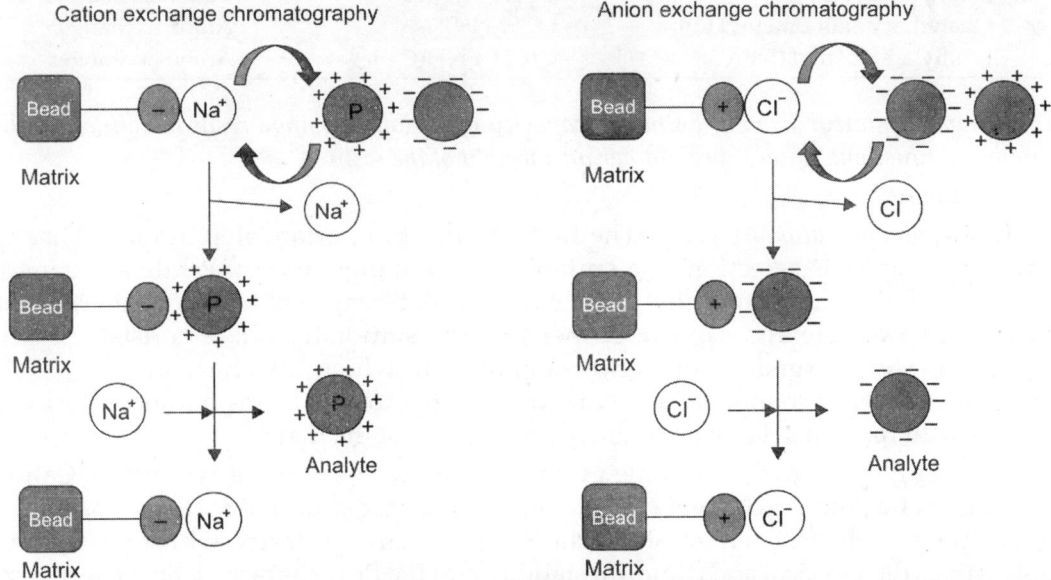

Fig. 6.5: Cation and anion exchange chromatography

Principle: The adsorbent used in ion-exchange chromatography is ionized resin matrix reversibly bound to oppositely charged ions. These reversibly bound ions get exchanged with the competitive ions present in analyte. The interaction between the resin matrix and the analyte is determined by the net charge, ionic strength and pH of the buffer. There are two types of ion-exchange chromatography:

Cation exchange chromatography: In this type of chromatography the matrix has negatively charged functional group reversibly bound to cations that are exchangeable with other similarly charged ions present in mobile phase. In presence of strong cation in the mobile phase, the positively charged ions bound to the matrix get replaced with the elution. The chemical reaction involved in cation exchange chromatography is as follows:

$$(Res\ A^-)\ B^+ + C^+ \rightarrow (Res\ A^-)\ C^+ + B^+$$

Anion exchange chromatography: The resin matrix used in anion exchange chromatography has positively charged functional groups bound to negatively charged molecules. During elution, the anions present in the analyte replace the reversibly bound negatively charged ions from the matrix. In the presence of a strong anion in the mobile phase, the matrix bound anions get replaced with elution. The reaction is given below.

$$(Res\ B^+)\ A^- + D^- \rightarrow (Res\ B^+)\ D^- + A^-$$

Some popular ion-exchange resins are listed in Table 6.1.

S. No.	Name	Functional group	Type of ion exchanger
	Table 6.1: Some popular ion-exchange resins with their functional group		
i.	Carboxylmethyl (CM)	$-OCH_2COOH$	Cation exchanger
ii.	Sulphonate (S)	$-OCH_2SO_3H$	Cation exchanger
iii.	Sulphopropyl (SP)	$-O(CH_2)_3SO_3H$	Cation exchanger
iv.	Quaternary aminomethyl (Q)	$-OCH_2N(CH_3)_3$	Anion exchanger
v.	Diethylaminoethyl (DEAE)	$-O(CH_2)_2NH(C_2H_5)_2$	Anion exchanger

Exchange capacity: *The ion-exchange capacity of an ion-exchange resin is defined as the number of functional groups present per unit weight of the resin.*

Procedure:

(I) *Selection of stationary phase:* The first and most important step in ion-exchange chromatography is selection of a suitable ion-exchange matrix. While selecting a matrix the following factors should be considered: Strength of ion exchanger, linear flow rate of sample and sample properties. The stationary phase consists of two elements: The charged groups and the matrix to which the charged groups are attached. The ion exchangers are classified according to both by the nature of the ionic groups with fixed matrix material and by the nature of the matrix.

According to the nature of ionic groups, ion exchangers are of two types: Cation exchanger and anion exchanger (*vide supra*). The charged groups decide the specificity and strength of ion exchanger by their density and polarity while the matrix determines the physical and chemical stability and the flow characteristics. The matrix properties that influence the chromatographic separation are: Physical structure (beaded or fibrous), particle size, pore size and structure, nature of the surface (hydrophobic or hydrophilic) and swelling properties. Beaded matrix shows higher chromatographic resolution compared to fibrous ion exchangers. Ion exchangers can be classified as macroporous, microporous and non-porous (Fig. 6.6). High porosity offers large surface area and thereby high density of ionic groups providing increased efficiency. However, non-porous exchanger matrices are chosen in order to avoid diffusion effect. Particles of uniform size are preferred over heterogeneous materials in order to achieve good resolution and reasonable flow rates.

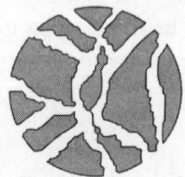 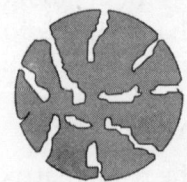 

Fig. 6.6: Different kind of matrices: Macroporous (left), microporous (middle), non-porous (right) resin beads

(II) *Packing the ion-exchange column:* An ion-exchange column has three parts: A long glass column, a valve assembly and a delivery tip. A small plug of glass wool is fixed at the bottom of the column to prevent the exchanger resin beads from blocking the valve assembly. The column is vertically placed with the help of a stand. Adequate amount of air-dried resin is taken in a beaker and slurry is made using DI water. Then the resin slurry is transferred completely into the column. One should be cautious while packing the column with slurry so that no air bubble is trapped inside the resin column. Once the column is set properly, it is washed with dil. HCl followed by DI water until the effluent gives negative result for Cl^- ion. The water level should always be kept 1 cm above the resin bed.

(III) *Elution and separation:* The liquid which enters the column is termed as *influent* and the liquid which comes out of the column is called *effluent*. The process of release of the absorbed ions is called *elution*. The solution used for elution is called *eluent* and the solution obtained from the column after elution is called *eluate*.

Cation exchange resin matrix contains H^+ ions as exchangeable ions attached with the anionic functional groups of the resin. When a solution containing mono-positive cations (say, Na^+) is passed through the column, it gets exchanged with the H^+ ions and the H^+ ion concentration in eluate is equivalent to the concentration of the cation in influent solution. The ion-exchange reaction of a sulfonated polystyrene-based cation exchanger with Na^+ ion is given below.

$$(Res.\ SO_3^-)H^+ + Na^+ \rightarrow (Res.\ SO_3^-)\ Na^+ + H^+$$

On the other hand, an anion exchanger exchanges OH^- ions with other exchangeable anions present in influent solution and the effluent contains equivalent amount of OH^- ions. A typical example of an anion exchange process is given in the following reaction.

$$[Res.\ N(Et)_3^+]OH^- + Cl^- \rightarrow [Res.\ N(Et)_3^+]Cl^- + OH^-$$

(IV) *Analysis of the eluate:* The eluate solution coming out from the column is collected in conical flasks. The eluate is then analysed for determination of H^+ or OH^- ion concentration by different techniques such as acid–base titration, spectrophotometric method (absorption measurement), and conductivity measurement.

Application of ion-exchange chromatography:

• *Softening of water:* The ion-exchange chromatography is mostly used in water treatment. This technique is very useful in removing excess Ca^{2+} and Mg^{2+} ions in hard water by replacing them with Na^+ ions of cation exchanger resins. After treatment exchanger is regenerated using high concentration solution of Na^+ ions.

• *Purification of water:* This chromatography technique is also useful in removing toxic heavy metal ions such as Pb^{2+}, Cd^{2+}, etc. Both cationic and anionic impurities can be removed using mixed bed resins along with application of alternating regeneration cycles.

• *Purification of metal:* The ion-exchange chromatography is very common method for purification of uranium from plutonium. Ion-exchange in combination with solvent extraction is routinely used for separation of lanthanides and actinides. It is also used in processing nuclear fuels and reprocessing of radioactive waste materials.

• Ion exchangers are also employed in purification of various pharmaceuticals.

• Ion exchangers are very popular in fruits and beverage industry. These exchangers are used to remove undesired compounds and thereby improve the taste and flavour of the product.

Experiment 6.1

Aim: Determination of ion-exchange capacity of cation exchange resin.

Principle: Ion exchange is a process which involves a rapid reversible exchange of similar charged ions between an insoluble solid phase (polymeric ion exchange resin) and a solution phase without any permanent change in structure. The ion-exchange resins are high molecular weight, crosslinked organic polymers containing charged groups having capability of capturing cations (cation exchange resin) or anions (anion exchange resin). It is used to demineralize water and to separate mixtures of ions. The quantity of ions loaded per unit weight of resin is called **ion-exchange capacity** of the resin which is expressed as milli-equivalents of exchangeable ion per gram (meq/g) of the resin.

The exchange capacity of a cation exchange resin can be expressed as the number of milli-equivalents of Na^+ ion adsorbed per gram of dry resin (hydrogen form). As Na^+ ions will replace equivalent number of H^+ ions, the exchange capacity can be estimated by determining the H^+ ion concentrations in the effluent.

An aqueous solution of sodium sulphate (Na_2SO_4) is passed through the resin column and the effluent is collected. This effluent contains H^+ ion in equivalent amount of Na^+ ion adsorbed by the resin. The exchange reaction can be expressed as:

$$2R\text{--}SO_3^- H^+ + 2Na^+ \text{ (aq)} + SO_4^{2-} \text{ (aq)} \rightarrow 2RSO_3^- Na^+ + 2H^+ \text{ (aq)} + SO_4^{2-} \text{ (aq)}$$

where, R represents resin unit.

The effluent is then titrated against standard NaOH solution using phenolphthalein indicator to determine the concentration of H^+ ion.

Apparatus: Column, stand and clamp, pipette, burette, conical flask, beaker, volumetric flask.

Chemical reagents:

(i) *Cation exchange resin in the hydrogen form*

(ii) *0.5 M Na_2SO_4 solution:* Dissolve 17.75 g of anhydrous Na_2SO_4 (MW = 142.04) in 250 mL DI water.

(iii) *0.1 N NaOH solution:* Dissolve 2.0 g NaOH in DI water and dilute to 500 mL in a volumetric flask.

(iv) *0.1 N oxalic acid solution:* Dissolve 3.15 g oxalic acid ($H_2C_2O_4$) in DI water and dilute to 500 mL in a volumetric flask.

(v) *Phenolphthalein indicator:* 0.5% solution in 1:1 alcohol.

Procedure:

(i) *Air-drying of resin:* Take the purified resin (hydrogen form) in evaporated dish and cover it with a watch glass. Keep the dish in warm condition (25–35 °C) overnight until it becomes free flowing.

(ii) *Setting the resin column:* The simple ion-exchange column has three parts—a long glass column, a valve assembly, and a delivery tip (shown in Fig. 6.7).

(a) Place a small plug of glass wool carefully at the bottom of the chromatography column. It prevents the beads of ion-exchange resin from blocking the valve assembly. Fill the column ~2/3rd full with DI water.

(b) Accurately weigh a small portion of the air-dried resin (~1 g) and take it in a 50 mL beaker. Add 25 mL DI water to it to make the slurry. Now, transfer the resin slurry into the column. To ensure complete transfer, wash the beaker for several times using DI water and transfer the washings into the column. Perform this step with great care so that no air bubble is trapped inside the resin column. After setting up the column adjust the water level to about 1 cm above

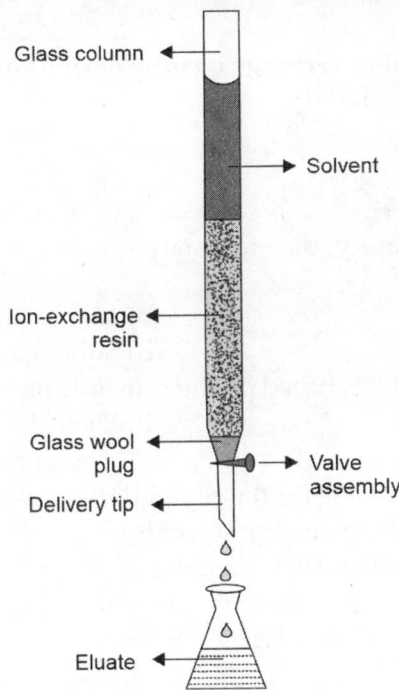

Fig. 6.7: Ion-exchange column

the surface of the resin bed. Never allow the water level to fall below the level of resin bed.

(iii) *Elution:* Elute the column with 150 mL of 0.5 M Na_2SO_4 solution in 5–10 mL portions at a rate of 2–3 mL/min. Collect the effluent in a 500 mL conical flask. After complete elution wash the resin bed 4–5 times with 5 mL portions of DI water and collect the washings in the same conical flask.

(iv) *Titration of the effluent:* Take the conical flask containing the effluent and add 1–2 drops of phenolphthalein indicator. Titrate the solution against standard 0.1 N NaOH solution until the end point is detected by a colour change from colourless to light pink.

(v) *Standardisation of NaOH solution:* Take out 25 mL portion of the standard 0.1 N oxalic acid in a 250 mL conical flask. Dilute the solution to 50 mL using DI water and add 1 drop of phenolphthalein indicator. Titrate it with NaOH solution until a light pink colour appears.

Calculation:

(i) Strength of NaOH solution (S)

$$= \frac{\text{Volume of oxalic acid (OA) solution} \times \text{strength of OA}}{\text{Volume of NaOH}}$$

(ii) *Determination of exchange capacity of resin:*

Weight of the cation resin taken (w) = ... g; Strength of NaOH (S) = ... N
Volume of NaOH required to titrate the effluent (V) = ... mL

Therefore, the ion-exchange capacity of the cation exchange resin $= \dfrac{V \times S}{w}$ meq/g

BIBLIOGRAPHY

1. Fisher S and Kunin R. Routine exchange capacity determinations of ion exchange resins. *Analytical Chemistry*, 1955, 27(7), 1191.

QUESTIONS

Multiple Choice Questions

1. In column chromatography, the stationary phase and the mobile phase are ...... and ......, respectively.
   - (a) Solid, liquid
   - (b) Liquid, liquid
   - (c) Liquid, gas
   - (d) Solid, gas
2. Column chromatography is based on the principle of:
   - (a) Ion-exchange
   - (b) Exclusion principle
   - (c) Differential adsorption
   - (d) Absorption
3. What is the factor responsible for the separation in column chromatography?
   - (a) Polarity differences between the solvents
   - (b) Polarity differences between the solutes
   - (c) No polarity difference between the solvents
   - (d) No polarity difference between the solutes
4. R_f is:
   - (a) Ruling factor
   - (b) Rolling factor
   - (c) Retardation factor
   - (d) None of these
5. The retention time depends on which of the following factor?
   - (a) Nature of the compound
   - (b) Column size and mode of packing
   - (c) Temperature
   - (d) All of the above
6. Which of the following factor affects the column efficiency?
   - (a) Dimension of the column
   - (b) Quality of the adsorbent
   - (c) Temperature
   - (d) All of the above
7. Ion-exchange chromatography is one kind of:
   - (a) Column chromatography
   - (b) Thin layer chromatography
   - (c) Paper chromatography
   - (d) None of these
8. The interaction between the solid phase and the analyte in ion-exchange chromatography is:
   - (a) Hydrogen bonding interaction
   - (b) van der Waals interaction
   - (c) Electrostatic interaction
   - (d) None of these
9. The solution used for analysis and the solution obtained from column are called ...... and ......, respectively.
   - (a) Eluate, eluent
   - (b) Eluent, eluate
   - (c) Eluent, electrolyte
   - (d) Electrolyte, eluate
10. The eluate from ion-exchange chromatography can be analysed by:
    - (a) Acid–base titration
    - (b) Conductivity measurement
    - (c) Spectrophotometric method
    - (d) All of the above

Answers

1. (a); 2. (c); 3. (b); 4. (c); 5. (d); 6. (d); 7. (a); 8. (c); 9. (b); 10. (a)

Practice Questions

1. What is column chromatography? What is the working principle for this kind of chromatography?
2. Briefly discuss the factors affecting column efficiency.
3. What is retention time? On which factors it depends?
4. Write down the identification procedure of different compounds separated by column chromatography.
5. Briefly discuss about the different classifications of column chromatography.
6. What is ion-exchange chromatography? Discuss briefly the working principle of ion-exchange chromatography.
7. Define exchange capacity. Write down the principle for determination of exchange capacity of an ion exchanger.
8. Write down different applications of ion-exchange chromatography.

7

Analysis of Cosmetics

UGC Syllabus

Major and minor constituents and their function:
1. Analysis of deodorants and antiperspirants: Al, Zn, boric acid, chloride, sulphate.
2. Determination of constituents of talcum powder: Magnesium oxide, calcium oxide, zinc oxide and calcium carbonate by complexometric titration

INTRODUCTION

The Section (aaa) of 'Drugs and Cosmetics Act, 1940' defines *"any article intended to be rubbed, poured, sprinkled or sprayed on, or introduced into, or otherwise applied to, the human body or any part thereof for cleansing, beautifying, promoting attractiveness, or altering the appearance and includes any article intended for use as a component of cosmetic."* So, the cosmetic is used to clean, protect, perfume and/or change the appearance of our body parts. There are thousands of various cosmetic items available in the market based on our demands, all with varying combinations of ingredients.

7.1 INGREDIENTS OF COSMETICS

India is one of the world's leading marketplace for cosmetics. There are various kinds of cosmetic products available in the market. In India, the cosmetic products are regulated under the 'Drugs and Cosmetics Act, 1940' and 'Drugs and Cosmetics Rules, 1945' and labelling declarations by the Bureau of Indian Standards (BIS).

A typical cosmetic product contains around 15 to 50 ingredients. These ingredients can be either naturally occurring or manmade chemicals. However, any potential impact of these on our health is mainly subject to the nature of the chemical compounds. The use of toxic chemicals should be in limited extent to prevent the adverse health effects in human. There are two Indian Standards namely, IS 4707 (Part 1): 2017 and 4707 (Part 2): 2017 describing colourants and impermissible chemicals, respectively. Although different product has different formulation, most of the cosmetics have a combination of at least some of the key or core ingredients. Among them water, thickener, emulsifier, emollient, colouring agent can be considered as the major constituents and preservative, fragrance, shimmers, etc. can be considered as minor constituents.

7.1.1 Major Constituents of Cosmetics

(I) *Cosmetics water:* If a cosmetic product is liquid or semi-liquid in nature such as makeup, lotions, creams, deodorants, conditioners, and shampoos, then the base of the product is most probably water. The water used in cosmetic is ultra-pure water which is free from microbes, toxins or other pollutants. Water often plays an important role in cosmetic processing by acting as a solvent for other ingredients forming emulsions for consistency.

(II) *Thickeners:* Thickeners are mainly used to enhance appealing consistency, volume and viscosity, as well as the stability and performance of the cosmetic products. Most of the thickeners have the capacity to hold water on its surface and thereby functions like moisturizers, while few of them act as emulsifiers or have gelling properties. Thickeners can be natural, synthetic or semi-synthetic.

Lipid thickeners are mainly used in lotions and creams. They primarily composed of lipophillic substances. These types of thickeners are generally solid in nature but liquefied while processing. They function by imparting their natural thickness to the formula. Some common lipid thickeners used in cosmetics are carnauba wax (queen of waxes), cetyl alcohol, stearic acid and stearyl alcohol.

Naturally derived thickeners are also used in various cosmetic products to increase the viscosity by adsorbing water. These types of thickeners are mainly naturally occurring polymers. Sometimes, their derivative is also used to improve its quality. Liquid cleansing products such as body wash or shampoo contains cellulose derivatives like hydroxyethyl cellulose as thickeners. Besides that xanthan gum, gelatin, guar gum, and locust bean gum are also very common naturally derived thickeners useful to decrease the level of water in cosmetic products.

Mineral thickeners are also naturally occurring materials. They not only absorb water but also oils to increase viscosity. However, this type of thickeners gives a different result than the natural gums. Some common mineral thickeners used in cosmetics are bentonite, magnesium-aluminium silicate, and silica.

Swellable polymer like carbomer is the most common synthetic thickener used in lotion and cream products. It is a nonlinear polymer of acrylic acid, cross-linked with a polyalkenyl polyether. Unlike other thickeners, they give pleasant feel even at high viscosities. However, they are very sensitive to pH and high ion content of the emulsion. Some other synthetic thickeners are cetyl palmitate, and ammonium acryloyldimethyltaurate.

(III) *Emulsifiers:* Emulsions are mixture of two or more immiscible liquids, in which one component is suspended or dispersed throughout another component as minute droplets. Many cosmetic products are based on either oil-in-water or water-in-oil emulsions. Since the two liquids are immiscible an emulsifier is required to prevent the substances (oil and water) from separating. Emulsifier produces a homogenous mixture with even texture by modifying the surface tension between the water and the oil. Cosmetic products like creams and lotions contain emulsifiers to mix water with oils. Some common emulsifiers used in cosmetics are laureth-4, polysorbates, PEG-100 stearate, potassium cetyl sulfate and distearyldimonium chloride.

(IV) *Emollient:* The word 'emollient' is actually derived from the Latin verb *'mollire'* means 'to soften'. In cosmetic industry, emollients are the ingredients introduced in a formulation which softens the skin by slowing evaporation of water. These are typically non-polar in nature and derived from both natural and synthetic sources. They are widely used in lipsticks, lotions and other cosmetics. Some common emollients are beeswax (Fig. 7.1), coconut oil, olive oil, lanolin, as well as mineral oil, petroleum jelly, zinc oxide, glycerin, diglycol laurate and butyl stearate.

Fig. 7.1: Beeswax used as emollient

(V) *Colouring agents:* Colouring agents are used in cosmetics in order to colour the product itself and/or change the colour of the skin. These are categorized in two main groups—colourants and pigments (Flowchart 7.1).

Colourants are soluble synthetic organic compounds which are used in various skin care products or toiletries. They can be obtained from natural sources (minerals, plants or animals) or synthetically produced. Depending on the solubility colourants can be classified into two types—hydrosoluble and liposoluble colourants. **Hydrosoluble colourants** consist of molecules containing one or more water-soluble groups like sulfonic ($-SO_3Na$) or carboxylic ($-COONa$). These colourants are very sensitive to UV rays, pH, and redox active chemicals. These are profoundly used in perfumes, emulsions, lotions, soaps and bath products. Caramel and carminic acid are examples of two widely employed hydrosoluble colourants. **Liposoluble colourants** are free from water-soluble groups and their oil stability limit is beyond a few grams per litre.

Flowchart 7.1: Schematic classification of colouring agents

These are UV sensitive and used in colouring anhydrous mixtures such as tanning oils, bath oils, sticks, etc. β-carotene is an example of liposoluble colourant.

Pigments are insoluble chemicals that remain as crystal or particles, and impart colour on the applied skin. Pigments can also be of two origins: Organic carbon-based molecules and inorganic or mineral which are generally metal oxides found as minerals. The three widely used organic pigments are lakes, toners and true pigments. The **lake pigments** are formulated by absorbing a dye into an insoluble metal salt such as alumina hydrate, titanium dioxide, aluminium benzoate, etc. This results in water insoluble lake pigments which find application in water-resistant or waterproof cosmetics. The **toner pigments** are water insoluble metal salts (generally calcium and barium salts) of organic dye molecules. The **true pigments** are the original dye compounds directly used in cosmetic formulations. Among **mineral pigments**, the most widely used are iron oxides (Fe_2O_3), zinc oxide (ZnO), chromium oxides (Cr_2O_3), titanium oxides (TiO_2), ultramarines (sodium aluminium sulfosilicates), and manganese violet ($MnNH_4P_2O_7$). Mineral pigments offer more heat and light resistant property than organic pigments. They are more opaque and less shiny but provide a longer-lasting colour.

7.1.2 Minor Constituents of Cosmetics

(I) *Preservatives:* Preservative is an important ingredient which has the anti-microbial activity. That means it inhibits the growth of microorganisms (Fig. 7.2) such as bacteria and fungi, which can decompose the cosmetic products. Depending upon the cosmetics formulation naturally occurring or synthetic preservatives are used with optimal amounts. Some of the common preservatives used in cosmetics are parabens, benzoic acids and salts, isothiazolinones, phenoxyethanol and formaldehyde releasers such as diazolidinyl urea.

(II) *Fragrances:* The most attractive ingredient of a cosmetic product is its fragrance or perfume. Fragrances are composed of natural (essential oils) or synthetic compounds (aroma) which are used in cosmetics not only to give the pleasant scent, sometimes it also mask the natural odour of other ingredients. More than 3000 compounds are utilized to synthesize large number of fragrances used in cosmetic

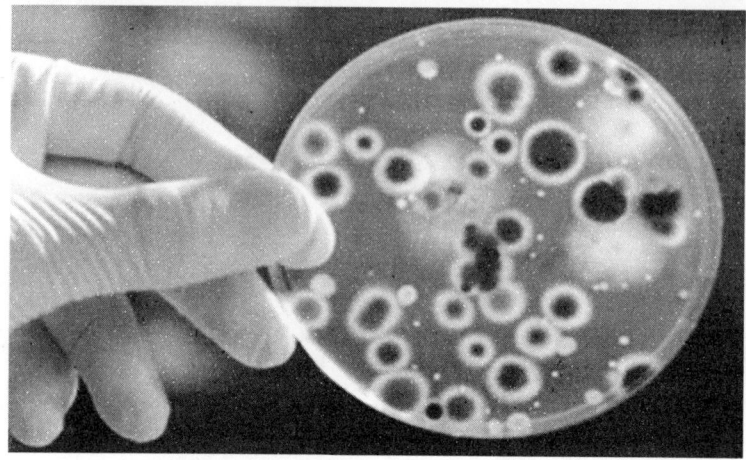

Fig. 7.2: Growth of microorganisms

products. All these compounds have to pass the safety standards of IFRA (International Fragrance Association) before being utilized as ingredients of fragrances.

7.2 ANTIPERSPIRANTS AND DEODORANTS

Antiperspirants and deodorants both are the personal care products which help to minimize the bad odour due to perspiration of our body but they both work in different paths.

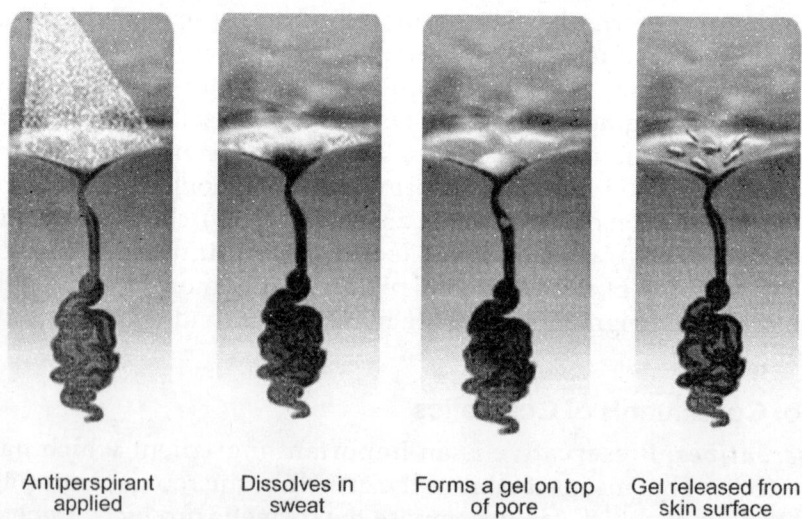

| Antiperspirant applied | Dissolves in sweat | Forms a gel on top of pore | Gel released from skin surface |

Fig. 7.3: Mechanism of antiperspirant

Antiperspirants control the humidity and odour in the axilla region by inhibiting perspiration. The mechanism of antiperspirant action is depicted in Fig. 7.3. The active ingredients are aluminium chloride, aluminium chlorohydrate and aluminium-zirconium tetrachlorohydrex glycine. The chemicals present in antiperspirants form a polymeric plug or blockage within the sweat duct that stops the flow of perspiration from the sweat glands to the skin surface. This plug is gradually destroyed over time, so reapplication is often needed.

The general preparation procedure of solid antiperspirants stick involves dispersion of waxes of fatty alcohols and couplers in silicone at high temperature. The working temperature is set according to the melting temperature of the castor wax to get a homogenous molten state. The fillers such as active powder, silica, talc, etc. are added at this step. Then the temperature is reduced to some extent, during this process the fragrance is added. Now, the temperature of the mixture is again reduced and poured into a container to get the final product. A cautious controlled over the cooling process is very essential for the final appearance of the antiperspirant stick. Before final cooling a light infrared heating can give a smooth surface of the stick. After 24–48 h of solidification homogeneity of the product is measured at various vertical depths by chemical analysis and hardness is measured using a penetrometer. A typical procedure for the preparation of an antiperspirant stick with formulation is given below (Table 7.1).

Table 7.1: General composition of an antiperspirant stick		
Part	Ingredient	Wt %
A	Decamethylcyclopentasiloxane	50.0
	Stearyl alcohol	13.5
	Hydrogenated castor oil (MP ~70°C)	3.5
	PPG 14 butyl ether	3.5
B	Aluminium-zirconium trichlorohydrex glycine (80% <10 mm)	24.0
	Talc	5.5
C	Fragrance	q.s.

q.s.: Quantum sufficit (a sufficient quantity)

Preparation of an antiperspirant stick: All the components of part A are gently mixed and the mixture is heated to 75 °C until it becomes clear. The mixture is then cooled down to 65 °C and a mixture of B is added and dispersed it to a homogenous mixture. Again lower the temperature is lowered to 56 °C. Then the fragrance (C) is added and the mixture is filled in the container at 54 °C.

Deodorants also help to minimize the body odour by antimicrobial action. The antibacterial compounds such as 2,4,4′-trichloro-2′-hydroxydiphenyl ether (triclosan), benzethoniumchloride, chlorhexidine acetate present in deodorants are responsible for the bacterial degradation in apocrine sweat secretions. In addition to the antimicrobial chemicals various fragrances are used in neat or in formulation as masking agents. However, none of these can provide long-term effect for extended time periods. Some metallic carbonates and zinc oxide also reveal deodorizing ability. A typical procedure for the preparation of a deodorant aerosol with formulation is given below (Table 7.2).

Table 7.2: General composition of a deodorant		
Part	Ingredient	Wt%
A	Aluminiumchlorohydrate powder	4.0
B	Dimethicone (50 cps)	5.0
C	C_{12}–C_{15} alcohol benzoate	15.0
D	Alcohol, SDA40	1.0
E	Organofunctional clay	0.8
F	Triclosan	0.1
G	Fragrance	q.s.
H	Isobutane/propane (80/20)	76.1

q.s.: Quantum sufficit (a sufficient quantity)

Preparation of deodorant aerosol: Using a high speed homogenizer ingredients B, C, E and F are mixed well to form dispersion. Then D and G are added with high speed continuous mixing which results in formation of a viscous gel. Small portions of A are then mixed to the above mixture slowly for homogenization. After complete addition, mixing is further continued for 15 min. After that the mixture is filtered through a 50 mesh sieve to aerosol containers and component H is mixed in 80:20 ratios.

7.3. ANALYSIS OF DEODORANTS AND ANTIPERSPIRANTS:
Al, Zn, Boric Acid, Chloride, Sulfate

Experiment 7.1

Aim: Determination of aluminium (Al) and zinc (Zn) in deodorants by gravimetric method.

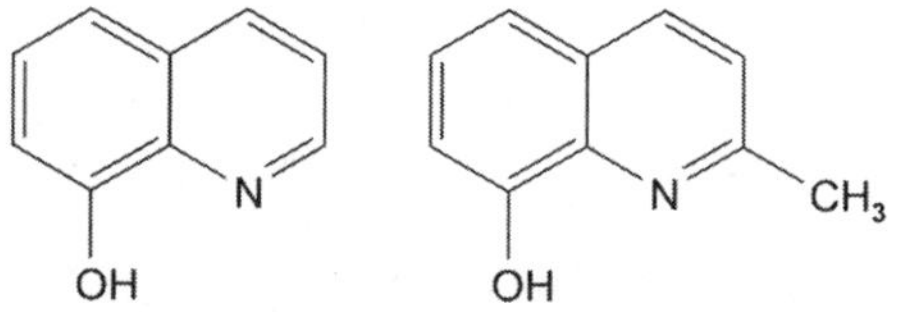

8-hydroxyquinoline   8-hydroxyquinaldine

Fig. 7.4: Molecular structures of 8-hydroxyquinoline and 8-hydroxyquinaldine

Principle: The total amount of aluminium and zinc is determined gravimetrically by the precipitation with 8-hydroxyquinoline. However, zinc is separately determined gravimetrically by precipitation with the more selective reagent, 8-hydroxyquinaldine.

Apparatus: Pipette, beaker, volumetric flask, filter paper, Gooch crucible.

Chemical reagents: (i) 2 N HCl, (ii) 2 N NH_4OH, (iii) 2 N NH_4OAc solution, (iv) phenolphthalein, (v) *8-hydroxyquinoline solution:* Dissolve 5 g of 8-hydroxyquino-line in 12 mL acetic acid and dilute to 100 mL with DI water and filter it. (vi) *8-hydroxy-quinaldine solution:* Dissolve 5 g of 8-hydroxyquinaldine in 12 mL acetic acid and dilute to 100 mL with DI water and filter it.

Sample preparation:

(I) *Liquid samples:* Take 5 mL of the sample (containing ~12–25 mg Al and ~20–60 mg Zn, see the manufacturer's label) in a 250 mL volumetric flask and dilute it to the mark with DI water. Take portion of this solution and filter it before analysis.

(II) *Solid samples:* Take 2–3 g of solid sample (containing ~12–25 mg Al and ~20–60 mg Zn, see the manufacturer's label) in a 100 mL beaker and add 5 mL of the 2 N HCl followed by addition of ~50 mL water. Boil the solution mixture for few minutes and after cooling filter it into a 250 mL volumetric flask. Wash the beaker for several times to ensure complete transfer and collect the washings into the same volumetric flask. Dilute the filtrate up to the mark of volumetric flask using DI water and make the solution homogeneous.

(III) *Cream and paste samples:* Take 2–3 g of sample (containing ~12–25 mg Al and ~20–60 mg Zn, see the manufacturer's label) in a 100 mL beaker and add 5 mL of the 2 N HCl followed by addition of ~50 mL water. Heat the mixture until the oils are liquefied and cool it down so that the oils are solidified. Collect the aqueous layer in a 250 mL volumetric flask through filtration. Smash the filter paper and return to the original beaker containing the solidified oils. Repeat the previous extraction procedure twice and each time collect the aqueous layer in the same volumetric flask. Finally wash the residue on the filter paper with DI water and collect the washings in the volumetric flask. Dilute the solution in the volumetric flask using DI water and make it homogeneous.

Procedure:

Step I: Precipitation of both Al and Zn as 8-hydroxyquinolato salt: Take 50 mL of sample solution (containing no interfering metals like Mg) in 400 mL beaker and add 2 drops of phenolphthalein. Then dropwise add the prepared 2 N NH_4OH solution until a faint permanent turbidity arises. Add 5 mL acetic acid and dilute the solution to 100 mL by adding DI water. Heat the mixture solution to 70–90 °C and add 10 mL of the 8-hydroxyquinoline solution at a time. After that, slowly add ammonium acetate solution to the reaction mixture. The final pH of the solution should be adjusted to 4.9–5.1 by addition of ammonium acetate for the yellow precipitation of the metal ion complexes. After complete precipitation, heat the reaction mixture below the boiling point for 2–5 min. Then allow the precipitate to settle down for 30–60 min. Filter the precipitate through Gooch crucible and wash thoroughly with DI water. Dry the precipitate at 130–140 °C for 1–2 h. Cool down to the room temperature and record the weight of the precipitate. Dry the precipitate again for 30 min, cool down to room temperature and weigh again. Repeat the procedure until a constant weight (W_1) is reached. This precipitate contains both tris(8-hydroxyquinolinato)aluminium complex $[Al(C_9H_6ON)_3]$ and bis(8-hydroxyquinolinato)zinc complex $[Zn(C_9H_6ON)_2]$.

Step II: Precipitation of Zn as 8-hydroxyquinaldinato salt: Take 50 mL of sample solution in 400 mL beaker and add 1 g ammonium tartrate, if Al present. Then add 2 mL 8-hydroxyquinaldine and dilute the solution using DI water and heat the mixture solution to 60–80 °C. Add NH_4OH dropwise to neutralize excess acid. Then slowly add 45 mL 2 N NH_4OAc solution with constant stirring. At this point, the pH of the solution should be ~5.5. Allow the solution to stand for about 20 minutes for the precipitation to complete. Filter the precipitate through Gooch crucible and wash thoroughly with DI water. Dry the precipitate at 130–140 °C for 1–2 h. Cool down to the room temperature and record the weight of the precipitate. Repeat heating, cooling and weighing until a constant weight (W_2) is reached. As the reagent selectively binds Zn^{2+}, this precipitate contains only bis(8-hydroxyquinaldinato) zinc complex $[Zn(C_{10}H_8ON)_2]$.

Calculation: The weight of the precipitate obtained in Step I (Al and Zn) = W_1 g and in Step II (Zn) = W_2 g.

1 mole of $[Zn(C_{10}H_8ON)_2]$ (MW = 381.74) contains 65.39 g Zn.

Therefore, amount of Zn in 50 mL sample extract $(x) = \dfrac{W_2 \times 65.39}{381.74}$ g.

This x g Zn is also present as $[Zn(C_9H_6ON)_2]$ (MW = 353.69) in Step I which weighs

$(W_{Zn}) = \dfrac{x \times 353.69}{65.39}$ g.

Hence, weight of $[Al(C_9H_6ON)_3]$ complex $(W_{Al}) = (W_1 - W_{Zn})$ g.

1 mole of $[Al(C_9H_6ON)_3]$ (MW = 459.43) contains 26.98 g Al.

Therefore, amount of Al in 50 mL sample extract $= \dfrac{W \times 26.98}{459.43}$ g

[NB: (i) If the cosmetic sample contains only Al, then amount of Al will be $\dfrac{W_1 \times 26.98}{459.43}$ g.

(ii) If the cosmetic sample contains only Zn, follow procedure of Step II only.]

Experiment 7.2

Aim: Determination of chloride (Cl^-) in deodorants by gravimetric method.

Principle: The total amount of chlorides is determined gravimetrically by precipitating it as AgCl using 0.1 N $AgNO_3$.

Apparatus: Beaker, volumetric flask, pipette, Gooch crucible, litmus paper.

Chemical reagents: (i) NH_4OH (1 : 1) solution, (ii) 0.01 N HNO_3, (iii) 0.1 N $AgNO_3$.

Sample preparation: See the sample preparation procedure in Experiment 7.1.

Procedure: Take 50 mL of sample solution (containing ~100 mg Cl^-, see the manufacture's label) in a 250 mL beaker and dilute it up to 150 mL by adding DI water. Add NH_4OH (1:1) solution until the sample solution becomes neutral (check with litmus paper) and then acidify the solution with 1 mL HNO_3 (1:1). If any undissolved precipitation remains, add little more HNO_3 to get a clear solution. Add slight excess ($\leq$5 mL) of 0.1 N $AgNO_3$ dropwise with constant stirring. White colour precipitate of AgCl starts to form immediately. Precipitation and succeeding work up is performed under low light. Heat the reaction mixture at 90–95 °C and stir until the precipitate coagulates. Keep the beaker undisturbed and let the precipitate settle down. Add 1–2 drops of 0.1 N $AgNO_3$ to the supernatant to ensure the presence of excess Ag and allow the mixture to stand for 1–2 h in dark.

Take a Gooch crucible and decant the supernatant first. Then wash the precipitate 2–3 times with 0.01 N HNO_3 and decant the washings through the Gooch crucible. Finally, transfer the precipitate with 0.01 N HNO_3 to the crucible. Continue washing the precipitate until the washings give negative test for Ag (no white precipitate is formed on addition of 1 drop 0.1 N HCl). Then wash the precipitate properly using DI water to remove the HNO_3 as much as possible. Heat the crucible for 1 h at 130–150 °C and weigh. Repeat drying until constant weight is reached.

Calculation: Weight of the Gooch crucible $= W_1$ g and the Gooch crucible containing AgCl precipitate $= W_2$ g.

$\therefore$ Weight of the AgCl precipitate $(W) = (W_2 - W_1)$ g

Hence, weight of chloride (Cl^-) $= = W \times \dfrac{35.45 \text{ (AW of chlorine)}}{143.32 \text{ (MW of AgCl)}}$ g present in 50 mL sample extract.

Experiment 7.3

Aim: Determination of sulfate (SO_4^{2-}) in deodorants by gravimetric method.

Principle: The total amount of sulfate is determined gravimetrically by precipitating $BaSO_4$ using 1% $BaCl_2$.

Apparatus: Beaker, litmus paper, Gooch crucible.

Chemical reagents: (i) NH_4OH (1:1) solution, (ii) dil. HCl, (iii) 1% $BaCl_2$ solution.

Sample preparation: See the sample preparation procedure in Experiment 7.1.

Procedure: Take 50 mL of sample solution (containing ~100 mg SO_4^{2-}, see the manufacture's label) in a 600 mL beaker and dilute it up to 350 mL by adding DI water. Add NH_4OH (1:1) solution until the sample solution becomes neutral (check with litmus paper) and then acidify the solution with 2 mL HCl. If any undissolved precipitation remains, add little more HCl to get a clear solution.

Heat the sample solution at near its boiling point. Take 50 mL 1% $BaCl_2$ solution in a 100 mL beaker and heat it at near boiling point temperature. Then rapidly add this 1% $BaCl_2$ solution to the sample solution. White colour precipitate of $BaSO_4$ starts to form immediately. Add slight excess of $BaCl_2$ solution with constant stirring to ensure the presence of excess Ba. Keep the beaker undisturbed on a steam bath for 1–2 h and let the precipitate settle down. Take a Gooch crucible and decant the supernatant first. Then wash the precipitate 4–5 times with warm DI water and decant the washings through the Gooch crucible. Finally transfer the precipitate with warm DI water to the crucible. Continue washing the precipitate until the washings give negative test for Cl (no white precipitate is formed on addition of few drops of 0.1 N $AgNO_3$). Heat the crucible for 2 h at 110–120 °C and weigh. Repeat drying until constant weight is reached.

Calculation: Weight of the Gooch crucible = W_1 g and the Gooch crucible containing $BaSO_4$ precipitate = W_2 g.

$\therefore$ Weight of the $BaSO_4$ precipitate $(W) = (W_2 - W_1)$ g

Hence, weight of sulfate (SO_4^{2-}) $= W \times \dfrac{96.06 \text{ (MW of sulfate)}}{233.38 \text{ (MW of } BaSO_4)}$ g present in 50 mL sample extract.

Experiment 7.4

Aim: Determination of boric acid (H_3BO_3) in deodorants and antiperspirants by ion-exchange method.

Principle: After removing the metal ions and organic part, boric acid is determined through titration method.

Apparatus: Chromatography column with stand, glass wool, beaker, porcelain casserole, round bottom (RB) flask, conical flask, distillation setup, quantitative filter paper, litmus paper.

Chemical reagents: (i) Amberlite IR 120(H) resin, (ii) dil. HCl (1:1), (iii) phenolph-thalein, (iv) NaOH solution 0.1 N and 10 % (w/v), (v) NH_4OH, (vi) methyl red indicator solution, (vii) mannitol.

Procedure:

(I) *Ion-exchange column preparation:* The simple ion-exchange column has three parts—a long glass column, a valve assembly, and a delivery tip.

   (a) Place a small plug of glass wool carefully at the bottom of the chromatography column. It prevents the beads of ion-exchange resin from blocking the valve assembly. Fill the column ~2/3rd full with DI water.

(b) Accurately weigh a small portion of the air-dried resin (~1 g) and take it in a 50 mL beaker. Add 25 mL DI water to it to make the slurry. Now, transfer the resin slurry into the column. To ensure complete transfer wash the beaker for several times using DI water and transfer the washings into the column. Perform this step with great care so that no air bubble is trapped inside the resin column. After setting up the column, wash with HCl and then with 50 mL H_2O until effluent gives negative test for chloride (Cl⁻) and finally adjust the water level to about 1 cm above the surface of the resin bed. Never allow the water level to fall below the level of resin bed.

(II) *Sample preparation: Removal of organic matter:* Take the sample 3–5 g (containing ~50–200 mg H_3BO_3, see the manufacture's label) in a 250 mL porcelain casserole and add 2 drops of phenolphthalein. Then add freshly prepared 10% NaOH solution to make it alkaline. Evaporate the solution to dryness on steam bath under gentle air current, and then dry the residue at 140°C for 1 h in oven. Convert the residue to ash by heating at 550 °C for 1 h. After cooling it to room temperature, add 50 mL hot DI water and then acidify it with HCl. Filter this hot solution through a quantitative filter paper into 250 mL beaker. Wash the filter paper with little amount of hot DI water and keep the filtrate (may be slightly cloudy) aside.

Transfer the filter paper to the same casserole and add 10 mL DI water followed by addition of few drops of 10 % NaOH until the solution becomes alkaline. Evaporate the solution to dryness on steam bath, and then dry the residue at 140°C for 1 h in oven. Convert the residue to ash by heating at 550 °C for 2 h. Cool and add 50 mL hot DI water and then acidify it with HCl. Filter this hot solution into the previous 250 mL beaker. Wash the casserole and filter paper thoroughly with hot DI water, and collect the washings in the same beaker.

Cool the combined filtrate solution and add NH_4OH until the solution becomes slightly alkaline (check with litmus paper) or flocculent precipitate appears. Re-acidify with HCl to make it slightly acidic or until precipitate just re-dissolves.

(III) *Elution:* Pass the above solution through the ion-exchange column carefully and collect the effluent into 1 L RB flask. Then wash the column with 50 mL portions of DI water for several times and collect the washings in the same RB flask. Add 5 mL of methyl red indicator and add freshly prepared 10% NaOH solution to make the solution alkaline, followed by HCl to make it barely acidic.

(IV) *Distillation and titration:* Connect the RB flask to a reflux condenser and boil for 5 min. Wash down the condenser with little DI water and cool the solution to room temperature. Neutralize the solution to methyl red with 0.1 M NaOH. Titrate the solution with 0.1 M NaOH in presence of 4–5 g mannitol and 0.5 mL phenolphthalein indicator solution. At the end point solution gives a pink colouration. Then add few more drops of mannitol, if pink colour disappears, continue titration until it reappears. Repeat addition of mannitol until a constant solution colour is achieved. Perform the titration in triplicate.

(V) *Determination of blank:* Take ~350 mL DI water in a 500 mL beaker and add volume of freshly prepared 10% NaOH solution equal to that required to neutralize the test sample after passing through the column. Acidify with HCl and proceed as above (distillation and titration procedure).

Calculation: Volume of NaOH required for the titration of test sample = v_1 mL and in blank titration = v_2 mL.

∴ Volume of NaOH required for the titration of H_3BO_3 (v) = $(v_1 - v_2)$ mL

1 mL 0.1 M NaOH ≡ 0.00618 g H_3BO_3 [MW of H_3BO_3 = 61.8]

Therefore, v mL 0.1 M NaOH = $(v \times 0.00618)$ g H_3BO_3.

7.4 DETERMINATION OF CONSTITUENTS OF TALCUM POWDER: MAGNESIUM OXIDE, CALCIUM OXIDE, ZINC OXIDE AND CALCIUM CARBONATE BY COMPLEXOMETRIC TITRATION

Talcum powder is one of the widely used cosmetic products. It is also known as dusting powder, body powder, bath powder or body talc. Talcum powder, often called talc, is mainly used to absorb moisture and sweat especially after bathing in warmer countries in order to keep the body cool. It has also lubricating property which helps in prevention of skin irritation due to chafing with clothes. Talcum powder can also be used as a substitute of perfumes.

Ingredients: The major ingredients of talcum powder is talc, which is finely powdered hydrous magnesium silicate having chemical formula, $[Mg_3(Si_4O_{10})(OH)_2]$ with a theoretical chemical composition of 31.7% MgO, 63.5% SiO_2 and 4.8% H_2O. The function of talc is to give slippery feel, so that the powder can be spread without dragging on the skin. However, commercial talc is rarely pure, which contains one or more foreign substances such as calcite, magnesite, kaolinite, quartz, chlorite, muscovite, serpentine, dolomite, tremolite, and asbestos (chrysotile). Even other materials are added to the talc to improve adhesion, absorbency, water repellence and add fragrance. Some other ingredients and their functions are listed in the Table 7.3.

Table 7.3: Other ingredients of talcum powder

Name	Functions
Zinc oxide (ZnO)	Astringent, smoothing agent, relieves prickling and irritation, sun-screening agent, bulking agent, skin protectant
Zinc stearate $[Zn(C_{18}H_{35}O_2)_2]$	Lubricating agent, antiseptic
Calcium carbonate ($CaCO_3$)	Fragrance carrier, abrasive, buffering agent, Bulking agent, opacifying agent
Magnesium carbonate ($MgCO_3$)	To increase fluffiness, and absorbing materials to carry perfume
Silica	Abrasive, absorbency
Boric acid (H_3BO_3)	Prevents bacterial growth, pH adjusters

Talcum powders from different commercial origin will have varying ratio of the constituent substances. It is not necessary that any talcum powder will contain all the above ingredients (mentioned in the Table 7.3) at a time. Formulation of some talcum powder is summarized in the Table 7.4.

Determination of various ingredients by complexometric titration: The amount of various ingredients such as MgO, ZnO, CaO, and $CaCO_3$ present along with talc in talcum powder is generally mentioned in the manufacturer's label. One can easily verify the exact amounts by complexometric titration of the metal ions using ethylenediaminetetraacetic acid (EDTA). By using suitable indicator (Table 7.5) one can easily perform the titration and calculate the exact amount of the ingredients.

Table 7.4: Formulation of some talcum powder				
Ingredients	I	II	III	IV
i. Zinc stearate	5	5	4	–
ii. Zinc oxide	–	5	–	4
iii. Calcium carbonate, light	25	–	25	8
iv. Magnesium carbonate, light	–	15	–	–
v. Boric acid	–	–	1	–
vi. Talc	70	75	70	88
vii. Perfumes	q.s.	q.s.	q.s.	q.s.

q.s.: Quantum sufficit (a sufficient quantity)

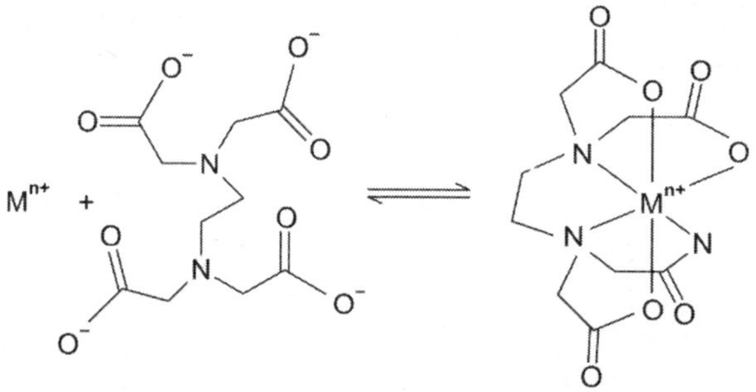

Fig. 7.5: Complexation of metal ion by EDTA

Complexometric titrations involve quantitative complexation of a metal ion by a suitable complexing agent. EDTA is the most widely used complexing agent due to some unique characteristics such as low cost, low toxicity, high stability or formation constant (K_f) with transition and alkaline earth metals in 1:1 stoichiometry, and control over selectivity with pH and masking agents. The stoichiometric reaction between metal ion and EDTA (Y^{4-}) is represented in Fig. 7.5. In most complexometric titrations, standard solution of EDTA is used as titrant. Numerous metallochromic or complexometric indicators are known in literatures (Table 7.5) that are employed to denote the end point. These indicators impart different colours when in free-state and complexed state. The preparation procedures of different buffer solutions used to maintain the pH of complexometric titrations are provided in Table 7.6.

• **Preparation of EDTA solution and its standardization:** 0.01 M EDTA solution: Take 3.72 g of ethylenediaminetetraacetic acid disodium salt dihydrate in a 400 mL beaker and add 200 mL DI water. After making a homogeneous solution transfer it to a 1000 mL graduated cylinder and dilute it with DI water up to the mark. Then keep the solution in a polyethylene flask.

Standardization of EDTA by CaCO₃: To prepare a standard calcium carbonate solution, weigh 0.3 g of the salt and dissolve it in 4 mL of 9% hydrochloric acid in a 250 mL volumetric flask. Then dilute the solution up to the mark with DI water. Take 15 mL aliquot of the standard calcium solution and add 80 mL of DI water followed by 4 pearls of KOH. Then add a pinch of solid 1% murexide indicator. Titrate the solution against 0.01 M EDTA until the colour changes from light pink to violet.

Calculate the strength of EDTA solution by using the following equation:

1 mL of 1 M CaCO₃ ≡ 1 mL of 1 M EDTA

Table 7.5: Some metal ion indicators and their preparations

Structure	Preparation
Eriochrome black T (EBT) indicator: 	1% (w/v): Dissolve 1 g of EBT in 100 mL 95% ethanol. *Solid mixture*: Mix 1 g EBT indicator with 100 g KNO_3 and grind the mixture well.
Murexide indicator: 	Solid mixture: Weigh 0.5 g of murexide and 49.5 g of NaCl. Mix the two solids and grind the mixture well. Finally, store the mixture in an opaque flask.
Xylenol orange indicator: 	0.2% (w/v): Dissolve 0.2 g of xylenol orange in 100 mL DI water.
Patton-Reeder indicator: 	0.2% (w/v): Dissolve 0.2 g Patton-Reeder reagent in methanol and dilute to 100 ml.
Calcon indicator: 	0.4% (w/v): Dissolve 0.4 g calcon in 100 mL of methanol.

Table 7.6: Some buffer solutions and their preparation

Buffer name	Preparation
Ammonia (NH_3–NH_4Cl) buffer (pH 10)	Dissolve 17.38 g of NH_4Cl in 143 mL of conc. NH_4OH (18.1 M, 35%) in a 400 mL beaker under fume-hood. Dilute to 250 mL with DI water, homogenize and store in an amber glass flask in the refrigerator.
Hexamine buffer (pH 5–6)	Dissolve 30 g hexamine in 70 mL DI water.

Experiment 7.5

Aim: Determination of magnesium oxide (MgO) in talcum powder by complexometric titration method.

Principle: The main ingredient of talcum powder, i.e. talc or hydrated magnesium silicate on fusion with anhydrous Na_2CO_3 gets converted to MgO. This MgO can then be measured by titration against standard EDTA solution.

Apparatus: Platinum crucible, glass-rod, beaker, filter paper, conical flask.

Reagent: (i) Anhydrous Na_2CO_3, (ii) conc. HCl, (iii) conc. ammonia solution, (iv) NH_3–NH_4Cl buffer (pH 10), (v) EBT indicator (solid mixture), (vi) 0.01 M EDTA solution

Sample preparation: Extraction of MgO from talcum powder: Take about 0.12 g of talcum powder in a dry platinum crucible and add about 0.6 g of anhydrous Na_2CO_3. Fuse the mixture by heating the crucible covered with a lid over a flame for about 30 min and extract the fused mass with portions of conc. HCl. Crush the mass with a glass-rod to help the extraction procedure. Transfer the extract quantitatively to a 150 mL beaker. Wash the crucible and the lid with DI water and transfer the washings to the same beaker. Dissolve by stirring and gentle crushing with glass rod. Boil the mixture solution, if required, for complete digestion. Cool down the solution to the room temperature and neutralize it by slow addition of ammonia solution till white precipitate appears. Add a little excess of ammonia and filter the mixture solution through a wet filter paper into a 500 mL conical flask. Wash the beaker with DI water for several times and collect the washings into the conical flask through filtration.

Procedure: Take the sample extract and add 10–12 mL NH_3–NH_4Cl buffer (pH 10) solution into the conical flask followed by the addition of a pinch EBT indicator and titrate the solution against standard 0.01 M EDTA to a red-blue end point. Perform the titration in triplicate.

Calculation: Amount of MgO can be calculated by using the following equation.

1 mL 1 M EDTA $\equiv$ 0.04030 of MgO

Experiment 7.6

Aim: Determination of zinc oxide (ZnO) in talcum powder by complexometric titration method.

Principle: Zinc oxide (ZnO), a white amorphous substance, has huge applications in cosmetic products mainly in talcum powder due to its antiseptic, anti-inflammatory, UV protection, deodorant, and skin healing properties. Several methods can be utilized for the determination of zinc, such as UV-vis spectrophotometry, flame atomic absorption spectrometry, and inductively coupled plasma mass spectrometry. However, titrimetric methods have some advantages over other methods due to their simplicity, high accessibility, low cost, and high accuracy in determining high concentrations of analyte.

The determination of ZnO present in the cosmetic sample is done by titrimetric method using ethylenediaminetetraacetic acid (EDTA). For this purpose ZnO is first extracted by dil. HCl and then titrated against EDTA using EBT indicator at pH 10. At above pH 4, the Zn–EDTA complex has a formation constant of ≥ 106. The free EBT indicator (HIn^{2-}) shows blue colour while the Zn–In complex is wine-red in colour. Initially, Zn^{2+} ions combine with EBT molecules to form Zn–In complex producing a wine-red colour solution. Near the end point, EDTA disrupts the Zn–In complexes to give Zn–EDTA complex and liberate free indicator which gives blue colour solution.

$$ZnO\,(s) + 2H^+\,(aq) \rightarrow Zn^{2+}\,(aq) + H_2O\,(l)$$

$$\underset{\substack{\text{Analyte}\\ \text{(colourless)}}}{Zn^{2+}} + \underset{\substack{\text{Free indicator}\\ \text{(blue)}}}{HIn^{2-}} \xrightleftharpoons{\text{pH 10}} \underset{\substack{\text{Metal ion–EDTA}\\ \text{complex}\\ \text{(wine red)}}}{ZnIn^-} + H^+$$

$$\underset{\substack{\text{Metal ion indicator}\\ \text{(wine red)}}}{ZnIn^-} + \underset{\substack{\text{Titrant}\\ \text{(EDTA)}}}{HY^{3-}} \xrightleftharpoons{\text{pH 10}} \underset{\substack{\text{Metal ion–EDTA}\\ \text{complex}\\ \text{(colourless)}}}{ZnY^{2-}} + \underset{\substack{\text{Free indicator}\\ \text{(blue)}}}{HIn^{2-}}$$

Apparatus: Beaker, glass rod, volumetric flask, filter paper, pipette, conical flask, burette.

Chemical reagents: (i) 9% HCl, (ii) NH_3–NH_4Cl buffer (pH 10), (iii) EBT indicator, (iv) 0.01 M standard EDTA solution.

Sample preparation: Extraction of ZnO from talcum powder: As ZnO is amphoteric in nature; it can be treated with both acids and bases. Take about 1.0 g of talc powder in a 250 mL beaker. Then add 50 mL of 9% HCl and stir the mixture using glass rod for 10 min. Transfer the solution quantitatively into a 250 mL volumetric flask and diluted up to the mark using DI water. Finally, filter the suspension into a 400 mL beaker. The clear filtrate solution contains Zn^{2+} ions extracted from the sample.

Procedure: Pipette out 20 mL of the sample extract containing Zn^{2+} in a 250 mL conical flask, to which add 10 mL of NH_3–NH_4Cl buffer of pH ~10. Then add two drops of EBT indicator. At this point, the solution will be magenta in colour. The mixture is then titrated with a 0.01 M standard solution of EDTA until a blue colour appears at the end point. Perform the titration in triplicate. Record the volume of EDTA consumed in the titration procedure.

Calculation: Amount of ZnO can be calculated by using the following equation:

$$1\text{ mL of }1\text{ M EDTA} \equiv 0.08138\text{ g of ZnO}$$

Experiment 7.7

Aim: Determination of calcium carbonate ($CaCO_3$) in talcum powder by complexo-metric titration method.

Principle: As calcium forms a relatively stable complex with EDTA (H_2Y^{2-}), it can be determined quantitatively by titration with EDTA.

$$Ca^{2+} + H_2Y^{2-} \longleftrightarrow CaY^{2-} + 2H^+$$

But when sample solution contains only Ca^{2+} ions, no sharp end point is detected, because the Ca–indicator complex formed with EBT is very weak and premature dissociation of the complex results in a colour change (red to blue) before the actual end point.

$$Mg^{2+} + H_2Y^{2-} \longleftrightarrow MgY^{2-} + 2H^+$$

On the other hand, Mg^{2+} ions form relatively less stable complex with EDTA, whereas the Mg–In complex is of intermediate stability between Ca–In complex and Mg–EDTA complex. Thus, the relative stability order of the complexes is as follows:

$$[Ca\text{–}EDTA] > [Mg\text{–}EDTA] > [Mg\text{–}In] > [Ca\text{–}In]$$

Therefore, while titrating a solution having both Ca^{2+} and Mg^{2+} ions with EDTA using EBT indicator, the EDTA first forms complex with free Ca^{2+} ions, then with free Mg^{2+} ions, and finally reacts with the Mg–In complex. The third reaction results in a colour change from wine-red (Mg–In complex) to blue (free indicator, In) which denotes the end point.

$$Mg\text{–}In + H_2Y^{2-} \longleftrightarrow MgY^{2-} + In$$

Hence, complexometric estimation of Ca^{2+} ions requires the presence of Mg^{2+} ions also. If Mg^{2+} ions are not present in the sample solution they need to be added in order to visualize the colour change of the indicator at actual end point. This is generally done by introducing a small amount of Mg–EDTA complex (1–10%) in the sample solution or in the buffer solution.

$$Ca^{2+} + MgY^{2-} \longleftrightarrow Mg^{2+} + CaY^{2-}$$

Apparatus: Beaker, glass rod, volumetric flask, filter paper, conical flask, pipette, burette, filter paper, pH paper.

Chemical reagents: (i) Dil. HCl (2 M), (ii) Dil. NaOH solution (2 M), (iii) 0.01 M standard EDTA solution, (iv) EBT indicator (solid mixture), (v) NH_3–NH_4Cl buffer solution (pH 10), (vi) 0.2 M solution of EDTA, (vii) phenolphthalein, (viii) *0.2 M $MgSO_4$ solution*: Dissolve 2.40 g $MgSO_4$ in 100 mL DI water in a volumetric flask, (ix) 0.01 M *$MgSO_4$ solution*: Dissolve 0.12 g $MgSO_4$ in 100 mL DI water in a volumetric flask, (x) *0.1 M Mg–EDTA complex:* Prepare the complex by mixing equal volumes of 0.2 M solution of EDTA and $MgSO_4$, followed by neutralization with NaOH to a pH 8–9 (phenolphthalein just gives red colour). For this complexation reaction one should be very careful about the equimolarity condition. To check this, take a small volume of the solution, add 1–2 drops of pH 10 buffer followed by a pinch of EBT indicator. The solution would be violet in colour. If the solution becomes blue on addition of one drop EDTA (0.01 M) and gives red colour solution on addition of one drop of 0.01 M $MgSO_4$ solution, the equimolarity of Mg^{2+} and EDTA is confirmed.

Sample preparation: Weigh ~0.5 g of the talc sample into a 100 mL beaker and add 20 mL dil. HCl. Stir the solution with a glass rod and allow the solid to dissolve completely. Neutralise the excess acid with dil. NaOH solution until pH 7 (check with pH paper). Transfer the solution quantitatively to a 100 mL volumetric flask via filtration and dilute it to the mark with DI water.

Procedure: Titration: Pipette out 25 mL of the sample solution into a 250 mL conical flask. Dilute it to 50 mL with DI water. Add 2 mL buffer solution followed by 1 mL 0.1 M Mg–EDTA complex and 30–40 mg EBT indicator. Titrate the solution against standard 0.01 M EDTA solution until the colour changes from wine-red to blue. Perform the titration in triplicate.

Calculation: Calculate the amount of $CaCO_3$ in the given talc sample using the following equation.

1 mL of 1 M EDTA $\equiv$ 0.10008 g of $CaCO_3$

Experiment 7.8

Aim: Determination of Ca^{2+} (CaO and/or $CaCO_3$) and Zn^{2+} (ZnO) in talcum powder by complexometric titration method.

Principle: To determine the amounts of Ca^{2+} and Zn^{2+} ions present in a solution mixture, firstly a portion of the solution mixture is titrated against standard EDTA using EBT indicator at pH 10, which gives the amount of total ions (Ca^{2+} and Zn^{2+}) present in that mixture. Then another portion (equal volume) of the mixture solution is again titrated with standard EDTA solution using xylenol orange indicator at pH 5–6, which gives only the concentration of Zn^{2+} ions present in the solution mixture. From the difference, the concentration of Ca^{2+} ions present in the solution mixture is calculated. Ca^{2+} ions can also be estimated in presence of Zn^{2+} ions by using KCN as masking agent for the Zn^{2+} ions.

Apparatus: Beaker, glass rod, volumetric flask, filter paper, conical flask, pipette, burette, glass rod, pH paper.

Reagent: (i) Dil. HCl (2 M), (ii) dil. NaOH solution (2 M), (iii) NH_3–NH_4Cl buffer solution (pH 10), (iv) EBT indicator, (v) 0.01 M EDTA solution, (vi) hexamine buffer (pH 5–6), (vii) xylenol orange indicator, (viii) triethanol amine, (ix) *20% KOH solution:* Dissolve 20 g of KOH in 100 mL DI water and mix it well. (x) *10% KCN solution:* Dissolve 1 g of KCN in 10 mL DI water and make the solution homogenous. (xi) Patton-Reeder indicator.

Sample preparation: See the sample preparation procedure in Experiment 7.7.

Procedure: Determination of total Ca^{2+} and Zn^{2+}: Pipette out 25 mL of the solution containing both Ca^{2+} and Zn^{2+} ions into a 250 mL conical flask and dilute it to 50 mL with DI water. Then add 5 mL of the NH_3–NH_4Cl buffer (pH 10) solution and mix it well. Add 50 mg of the EBT indicator. Titrate the solution against standard EDTA solution until the colour changes from wine-red to pure blue. Perform the titration in triplicate. Record the average volume of the EDTA solution required in this step (say, V_1 mL).

Determination of Zn^{2+}: Pipette out 25 mL aliquot from the sample extract in a 250 mL conical flask. Add 7 mL hexamine buffer (pH 5 to 6) and few drops of xylenol orange indicator. Titrate the solution against standard EDTA until the colour changes from orange-red to yellow. Perform the titration in triplicate. Record the average volume of the EDTA solution required in this step (say, V_2 mL).

Determination of Ca^{2+}: Pipette out 25 mL aliquot from the sample extract in a 250 mL conical flask and add 2–3 drops of triethanol amine under stirring condition. Add 50 mL of 20% KOH solution followed by addition of 0.5–1.0 mL of 10% KCN solution to mask Zn^{2+} ions. Then add 5–6 drops of Patton-Reeder indicator. At this point, the solution colour will be rose-red. Titrate the solution against standard EDTA until the colour changes from rose red to blue. Perform the titration in triplicate. Record the average volume of the EDTA solution required in this step (say, V_3 mL).

[NB: Potassium cyanide (KCN) is very much toxic and should be handled with extreme care. However, only Ca^{2+} can be determined from the value of ($V_1 - V_2$) and the experiment (determination of Ca^{2+}) which includes KCN as masking agent can be avoided.]

Calculation: Volume of EDTA required for Ca^{2+} = ($V_1 - V_2$) mL

Volume of EDTA required for Zn^{2+} = V_2 mL

Amount of CaO and ZnO can be calculated from the following equations:

1 mL of 1 M EDTA ≡ 0.05608 g of CaO

1 mL of 1 M EDTA ≡ 0.08138 g of ZnO

Experiment 7.9

Aim: Determination of Ca^{2+} (CaO and/or $CaCO_3$) and Mg^{2+} (MgO) in talcum powder by complexometric titration method.

Principle: The determination of only Ca^{2+} in a mixture of CaO and MgO can be performed by complexometric titration using calcon indicator. In this method, Mg^{2+} ions are precipitated quantitatively before titration of the solution. Titration using EBT indicator gives the total amount of Ca^{2+} and Mg^{2+} ions, and by subtracting the concentration of Ca^{2+} from the total amount, Mg^{2+} ions can be estimated.

Apparatus: Beaker, glass rod, volumetric flask, filter paper, conical flask, pipette, burette, pH paper.

Reagent: (i) Dil. HCl (2 M), (ii) dil. NaOH solution (2 M), (iii) NH_3–NH_4Cl buffer solution (pH 10), (iv) EBT indicator, (v) 0.01 M EDTA solution, (vi) diethylamine, (vii) calcon indicator.

Sample preparation: See the sample preparation procedure in Experiment 7.7.

Procedure: Determination of total Ca^{2+} and Mg^{2+}: Pipette out 25 mL of the solution containing both Ca^{2+} and Mg^{2+} ions into a 250 mL conical flask and dilute it to 50 mL with DI water. Then add 5 mL of the NH_3–NH_4Cl buffer (pH 10) solution and mix it well. Add 50 mg of the EBT indicator. Titrate the solution against standard EDTA solution until the colour changes from wine-red to pure blue. Perform the titration in triplicate. Record the average volume of the EDTA solution required in this step (say, V_1 mL).

Determination of only Ca^{2+}: Pipette out 50 mL of the sample extract and treat with 5 mL of diethylamine. At this point, the pH of the solution becomes 12.5 and Mg^{2+} ions are quantitatively precipitated as $Mg(OH)_2$. Then add 4–5 drops of calcon indicator to the solution. Titrate the solution against standard EDTA solution under stirring condition. At the end point colour of the solution changes from pink to pure blue. Perform the titration in triplicate. Record the average volume of the EDTA solution required in this step (say, V_2 mL).

Calculation: Volume of EDTA required for $Mg^{2+} = (V_1 - V_2)$ mL

Volume of EDTA required for $Ca^{2+} = V_2$ mL

Amount of CaO and MgO can be calculated from the following equations:

1 mL of 1 M EDTA $\equiv$ 0.05608 g of CaO

1 mL of 1 M EDTA $\equiv$ 0.04030 g of MgO

Experiment 7.10

Aim: Determination of Mg^{2+} ($MgCO_3$) and Zn^{2+} (ZnO) in talcum powder by complexometric titration method.

Principle: To determine the amounts of Mg^{2+} and Zn^{2+} ions present in a solution mixture, firstly a portion of the solution mixture is titrated against standard EDTA using EBT indicator at pH 10, which gives the amount of total ions (Mg^{2+} and Zn^{2+}) present in that mixture. Then another portion (equal volume) of the mixture solution is again titrated with standard EDTA solution using xylenol orange indicator at pH 5–6, which gives only the concentration of Zn^{2+} ions present in the solution mixture. By subtracting the concentration of Zn^{2+} ions from the total amount, the concentration of Mg^{2+} ions present in the solution mixture is calculated.

Apparatus: Beaker, glass rod, volumetric flask, filter paper, conical flask, pipette, burette, pH paper.

Reagent: (i) Dil. HCl (2 M), (ii) dil. NaOH solution (2 M), (iii) NH_3–NH_4Cl buffer (pH 10), (iv) EBT indicator, (v) standard EDTA solution, (vi) hexamine buffer (pH 5 to 6), (vii) xylenol orange indicator.

Sample preparation: See the sample preparation procedure in Experiment 7.7.

Procedure: Determination of total Mg^{2+} and Zn^{2+}: Pipette out 25 mL of the solution containing both Mg^{2+} and Zn^{2+} ions into a 250 mL conical flask and dilute it to 50 mL with DI water. Then add 5 mL of the NH_3–NH_4Cl buffer (pH 10) solution and mix it well. Add 50 mg of the EBT indicator. Titrate the solution against standard EDTA solution until the colour changes from wine-red to pure blue. Perform the titration in triplicate. Record the average volume of the EDTA solution required in this step (say, V_1 mL).

Determination of Zn^{2+}: Pipette out 25 mL aliquot from the sample extract in a 250 mL conical flask. Add 7 mL hexamine buffer (pH 5 to 6) and few drops of xylenol orange indicator. Titrate the solution against standard EDTA until the colour changes from orange-red to yellow. Perform the titration in triplicate. Record the average volume of the EDTA solution required in this step (say, V_2 mL).

Calculation: Volume of EDTA required for $Mg^{2+} = (V_1 - V_2)$ mL

Volume of EDTA required for $Zn^{2+} = V_2$ mL

Amount of MgO and ZnO can be calculated from the following equations:

1 mL of 1 M EDTA $\equiv$ 0.04030 g of MgO

1 mL of 1 M EDTA $\equiv$ 0.08138 g of ZnO

Experiment 7.11

Aim: Determination of Ca^{2+}, Mg^{2+} and Zn^{2+} in talcum powder by complexometric titration method.

Principle: To determine the amounts of Ca^{2+}, Mg^{2+} and Zn^{2+} ions present in a solution mixture, firstly a portion of the solution mixture is titrated against standard EDTA using EBT indicator at pH 10, which gives the amount of total ions (Ca^{2+}, Mg^{2+} and Zn^{2+}) present in that mixture. Then another portion (equal volume) of the mixture solution is again titrated with standard EDTA solution using xylenol indicator at pH 5–6, which gives only the concentration of Zn^{2+} ions present in the solution mixture. Then a third portion (equal volume) of the mixture solution is again titrated with standard EDTA solution using Patton-Reeder indicator after precipitating out $Mg(OH)_2$ and masking Zn^{2+} ion by KCN. This gives only the concentration of Ca^{2+} ions present in the solution mixture. From the difference, the concentration of Mg^{2+} ions present in the solution mixture is calculated.

Apparatus: Beaker, glass rod, volumetric flask, filter paper, conical flask, pipette, burette, glass rod, filter paper, pH paper.

Reagent: (i) Dil. HCl (2 M), (ii) dil. NaOH solution (2 M), (iii) NH_3–NH_4Cl buffer (pH 10), (iv) EBT indicator, (v) standard EDTA solution, (vi) hexamine buffer (pH 5 to 6), (vii) xylenol orange indicator, (viii) triethanol amine, (ix) *20% KOH solution:* Dissolve 20 g of KOH in 100 mL DI water and mix it well. (x) *10% KCN solution:* Dissolve 1 g of KCN in 10 mL DI water and make the solution homogenous. (xi) **Patton-Reeder indicator.**

Sample preparation: See the sample preparation procedure in Experiment 7.7.

Procedure: Determination of total Ca^{2+}, Mg^{2+} and Zn^{2+}: Pipette out 25 mL of the solution containing Ca^{2+}, Mg^{2+} and Zn^{2+} ions into a 250 mL conical flask and dilute it to 50 mL with DI water. Then add 5 mL of the NH_3–NH_4Cl buffer (pH 10) solution and mix it well. Add 50 mg of the EBT indicator. Titrate the solution against standard EDTA solution until the colour changes from wine-red to pure blue. Perform the titration in triplicate. Record the average volume of the EDTA solution required in this step (say, V_1 mL).

Determination of Zn^{2+}: Pipette out 25 mL aliquot from the sample extract in a 250 mL conical flask. Add 7 mL hexamine buffer (pH 5 to 6) and few drops of xylenol orange indicator. Titrate the solution against standard EDTA until the colour changes from orange-red to yellow. Perform the titration in triplicate. Record the average volume of the EDTA solution required in this step (say, V_2 mL).

Determination of Ca^{2+}: Pipette out 25 mL aliquot from the sample extract in a 250 mL conical flask and add 2–3 drops of triethanol amine under stirring condition. Add 50 mL of 20% KOH solution followed by addition of 0.5–1.0 mL of 10% KCN* solution to mask Zn^{2+} ions. Then add 5–6 drops of Patton-Reeder indicator. At this point, the solution colour will be rose-red. Titrate the solution against standard EDTA until the colour changes from rose-red to blue. Perform the titration in triplicate. Record the average volume of the EDTA solution required in this step (say, V_3 mL).

Calculation: Volume of EDTA required for $Mg^{2+} = (V_1 - V_2 - V_3)$ mL

Volume of EDTA required for $Zn^{2+} = V_2$ mL

Volume of EDTA required for $Ca^{2+} = V_3$ mL

Amount of CaO, MgO and ZnO can be calculated from the following equations:

1 mL of 1 M EDTA $\equiv 0.05608$ g of CaO

1 mL of 1 M EDTA $\equiv 0.04030$ g of MgO

1 mL of 1 M EDTA $\equiv 0.08138$ g of ZnO

BIBLIOGRAPHY

1. Butler H (ed.). *Poucher's perfumes, cosmetics and soaps.* 2013, Springer Science & Business Media.
2. Clements JE. Report on Deodorants and Anti-Perspirants. *Journal of Association of Official Agricultural Chemists*, 1953, 36(3), 791–793.
3. Jones JH. Report on Deodorants and Anti-Perspirants. *Journal of Association of Official Agricultural Chemists*, 1945, 28(4), 734–739.
4. Kramer H. Report on Deodorants and Anti-Perspirants. *Journal of Association of Official Agricultural Chemists*, 1950, 33(2), 371–374.
5. Mendham J. *Vogels textbook of quantitative chemical analysis.* 2006, Pearson Education India.
6. Merritt Jr LL and Walker JK. 8-Hydroxyquinaldine as Analytical Reagent. *Industrial & Engineering Chemistry Analytical Edition*, 1944, 16(6), 387–389.
7. Moyer HV and Remington WJ. Coprecipitation and pH Value in Precipitations with 8-Hydroxyquinoline. *Industrial & Engineering Chemistry Analytical Edition*, 1938, 10(4), 212–213.
8. The drugs and cosmetics act and rules, Ministry of health and family welfare, Government of India.
9. Williams SD and Schmitt WH. *Chemistry and technology of the cosmetics and toiletries industry*, 2012, Springer Science & Business Media.

---

* Potassium cyanide (KCN) is very much toxic and should be handled with extreme care.

QUESTIONS

Multiple Choice Questions

1. Which act is applicable to regulate the cosmetic products in India?
   (a) Drugs and Cosmetics Act 1940
   (b) Drugs and Cosmetics Rules 1945
   (c) Both a and b
   (d) None of these

2. Which of the following is not a major constituent of cosmetics?
   (a) Emulsifier
   (b) Emollient
   (c) Fragrance
   (d) Colouring agent

3. Emollients are used in cosmetics as:
   (a) Softener
   (b) Drying agent
   (c) Preservatives
   (d) None of these

4. Ultramarines are:
   (a) Hydrosoluble colourants
   (b) Inorganic pigments
   (c) Liposoluble colourants
   (d) Organic pigments

5. Antiperspirants control the body odour by:
   (a) Inhibiting perspiration
   (b) Minimizing antimicrobial action
   (c) Both a and b
   (d) None of these

6. For gravimetric analysis of Al and Zn using 8-hydroxyquinoline (A) and 8-hydroxyquinaldine (B), which of the following statements is correct?
   (a) A is more selective for Zn
   (b) A is more selective for Al
   (c) B is more selective for Zn
   (d) B is more selective for Al

7. Talc contains which of the following elements?
   (a) Zinc, calcium and oxygen
   (b) Calcium, oxygen and tin
   (c) Magnesium, silicon, oxygen
   (d) Zinc, tin and sulphur

8. Which of the following ingredient in talc acts as lubricating agent?
   (a) Zinc oxide
   (b) Zinc stearate
   (c) Silica
   (d) Calcium carbonate

9. Calcon and xylenol orange indicators are selectively used in complexometric titration of ...... and ...... ions, respectively.
   (a) Ca, Zn
   (b) Zn, Ca
   (c) Mg, Ca
   (d) Ca, Mg

10. While analysing a mixture of Ca, Mg and Zn, KCN is used as:
    (a) Chelating agent
    (b) Masking agent
    (c) Indicator
    (d) Buffering agent

Answers

1. (c); 2. (c); 3. (a); 4. (b); 5. (a); 6. (c); 7. (c); 8. (b); 9. (a); 10. (b)

Practice Questions

1. Define cosmetics. What are the major and minor constituents of cosmetics?
2. Briefly discuss about different kinds of thickeners used in cosmetics.
3. Classify the colouring agents based on their solubility with brief discussion.

4. What is the role of preservatives in cosmetics?

   [Hint: Preservatives are used in order to retard the degradation of a cosmetic product by microorganisms during formulation, storage, shipment and customer use. For example, antioxidants are useful for protection of the products from oxygen.]

5. How do antiperspirants and deodorants work?

6. Does antiperspirants possess any harmful effect by inhibiting perspiration?

   [Hint: Antiperspirant does not interfere in thermoregulation of our body and it is tested to be safe for being used in cosmetic products. The underarm areas only cover 1% (200 cm^2) of our body which is very small to cause any effect on thermoregulation process.]

7. Write down the principle of gravimetric analysis of Al and Zn in deodorants. Why 8-hydroxyquinaldine is a more selective reagent than 8-hydroxyquinoline?

   [Hint: Although 8-hydroxyquinoline and 8-hydroxyquinaldine closely resemble each other in chemical functionality, the latter is more selective in nature. Due to increased size of 8-hydroxyquinaldine, it is difficult for the ligand to form the octahedral donor cage around a small ion by assembling three large ligand molecules. This is the reason that 8-hydroxyquinaldine does not form complex with Al^{3+} ion with which 8-hydroxyquinoline reacts efficiently. On the other hand 8-hydroxyquinaldine forms complexes with Fe^{3+} ions very easily even in less acidic solutions.]

8. What is talcum powder? What are the major ingredients of talcum powder? Discuss about their functions.

9. What is the purpose of using talcum powder?

   [Hint: Talcum powder is primarily used to absorb moisture. It can be used as personal hygiene product, cosmetics, for preventing rash, etc.]

10. Write a short note on complexometric titration.

11. Why a polythene bottle is used to store EDTA?

    [Hint: Glass contains large number of metals ions. If one keeps EDTA in a glass container, it can form complexes with the metal ions. Thus, the concentration of EDTA solution will be changed and it will be contaminated.]

12. What do you mean by metallochromic indicator? Give two examples.

8

Analytical Chemistry in Forensic Sciences

INTRODUCTION

Analytical chemistry is a branch of chemistry where different methods and instruments are used to analyse an unknown chemical substance qualitatively as well as quantitatively. With advancements in this subdivision of chemistry, analytical chemistry has become an important tool of investigation in forensic sciences. Forensic science deals with trace analysis, drug analysis, arson analysis, toxin analysis, etc. in investigating a crime scene and analysis of evidences obtained from there with proper analytical techniques. Following three experiments of analytical chemistry methods, often used in forensic analysis have been included in the new CBCS syllabus and thereof discussed in this chapter:
  (i) Uses of phenolphthalein in trap cases
 (ii) Analysis of arson accelerants and
(iii) Analysis of gasoline sample

8.1 USES OF PHENOLPHTHALEIN IN TRAP CASES

In most of the bribe trap cases phenolphthalein is used as trapping agent. Phenolphthalein is a fine colourless chemical powder which is spread on the currency notes that are going to be handed over for the bribery. Locard's exchange principle states that every contact leaves a trace. So, when the accused person takes the currency notes, the phenolphthalein powder gets transferred from currency notes to his hands or other objects like his clothes, bag, pockets, drawer, file, table, and briefcase whichever come in contact with. Then his hands or other objects are washed with a colourless 2–3% sodium carbonate solution, which becomes immediately pink colour. Finally, these washings are referred to the forensic chemical laboratories along with other related substances for complete analysis using TLC, spectrophotometry, HPLC, FTIR, and GC-MS, etc. (Fig. 8.1).

Chemistry of phenolphthalein: Phenolphthalein is a white coloured organic compound having chemical formula, $C_{20}H_{14}O_4$ and often represented as H_2In. It is extensively used as indicator in acid–base titration. Depending on the pH of the medium, phenolphthalein can exist in either protonated or deprotonated form. It can

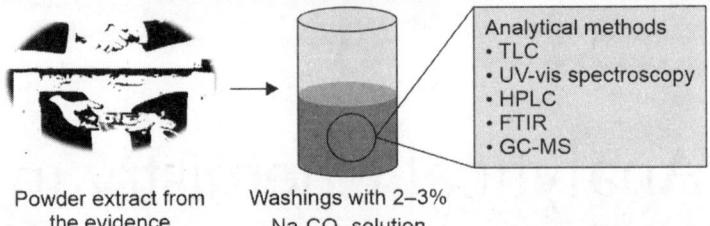

Fig. 8.1 Different analytical tools for detection of phenolphthalein

| | H₃In⁺ | H₂In | In²⁻ | In(OH)³⁻ |

pH: <0   0.0 – 8.2   8.2–12.0   >12.0
        Strongly acidic   Acidic or ~neutral   Basic   Strongly basic
Colour: Orange   Colourless   Pink   Colourless

Fig. 8.2: pH dependence of the colour of phenolphthalein

possess four different forms H₃In⁺ (pH <0, orange colour), H₂In (pH = 0–8.2; colourless), In²⁻ (pH = 8.2–12.0; pink) and In(OH)³⁻ (pH >12.0; colourless) (Fig. 8.2).

Extraction of phenolphthalein: Collect an appropriate amount of washing solution in a 100 mL beaker and acidify it with dil. HCl until the pH value of ~ 4–5 is reached. Then extract the solution three times with 30 mL diethylether. The volume of the ether extract is reduced and the solute is crystallized as white crystalline mass (MP 213 °C). The obtained residue is used for colour tests, thin layer chromatography, spectrophotometric analysis, etc.

Experiment 8.1

Aim: Detection of phenolphthalein by thin layer chromatography (TLC).

Principle: Thin layer chromatography (TLC) is a technique for analysing mixtures by separating the different components present in the mixture. This method includes three steps: Spotting, development, and visualization. For identification of a particular substance in a given sample, often a reference spotting is done where commercially available pure chemical substances are used. In trap case analysis of the washings, the residue obtained from the extract is spotted along with pure phenolphthalein spot on the TLC plate.

Apparatus: Beaker, capillary tube, TLC chamber.

Reagents:

   Stationary phase: Silica Gel G
   Mobile phase:
   (i) Chloroform : Acetone (4 : 1), or
   (ii) Benzene : Dioxane : Acetic acid (15 : 3 : 2), or
   (iii) Ethyl acetate : Methanol : Ammonia (17 : 2 : 1)

Procedure: Firstly, spot the ether extract of the washing solution and standard phenolphthalein as reference on silica gel plate. Then for developing, keep the silica gel plate in one of the solvent system (mobile phase) which is then moved up on the plate by capillary action. When the solvent has moved nearly to the top of the plate, take the plate out and allow the solvent to evaporate. The spots can be visualized in either of the two methods described below:

(i) Expose the TLC plate to ammonia vapour or iodine fumes for visual development of undecomposed phenolphthalein pink spot and compare with the standards.

(ii) Spray the TLC plate with dil. NaOH followed by acidified $KMnO_4$ solution. After that, hold the TLC plate under UV light to visualize fluorescent spots of phenolphthalein.

Experiment 8.2

Aim: Detection of phenolphthalein by spectrophotometric method.

Principle: Phenolphthalein can exist in two forms—acid form (H_2In) and base form (In^{2-}) depending on the pH of the medium. These two forms (H_2In and In^{2-}) correspond to the two absorption bands, respectively, resulting from the π-π^* and n-π^* transitions. Ethanol solution of phenolphthalein at lower pH (3–8) produces one absorption band at ~280–350 nm whereas at higher pH (9–12) an additional absorption peak at ~555 nm is observed (Fig. 8.3). The presence of traces of phenolphthalein in trap case washing extract is confirmed by recording its absorption spectra and comparing the same with that of pure phenolphthalein at the same pH.

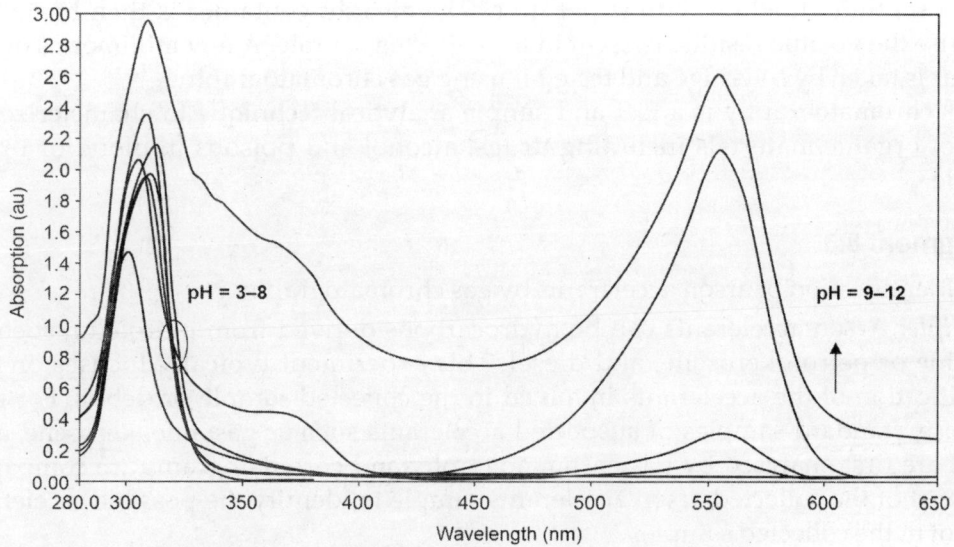

Fig. 8.3: UV-vis spectra of phenolphthalein at different pH

Apparatus: UV-vis spectrophotometer, cuvette, pipette, beaker.

Chemical reagents: Dil. NaOH solution.

Procedure: Take a small portion of the extract solution in a cuvette and dilute properly. Record the absorbance spectrum using a spectrophotometer. The UV spectrum records an absorbance maximum in the range of 280–350 nm. Take the cuvette out and add 2 drops of dilute NaOH solution and make the solution homogeneous. Again record the UV spectrum. A new absorbance maximum is observed in the range of 550–555 nm.

Other instrumental methods for detection of phenolphthalein are HPLC, FTIR, and GC-MS.

8.2 ANALYSIS OF ARSON ACCELERANTS

Arson is the criminal act of deliberately setting fire to a property claiming lives of a large number of people as well as wealth on this earth. Fire originates from a point of source and moves upward and outward simultaneously. If a liquid accelerant is used in such a case, a small portion of the accelerant can be soaked by the downwards portion. To investigate the origin or point of the fires, the investigators try to collect the samples by digging downward

Fig. 8.4: Collection of sample for arson accelerants analysis

(Fig. 8.4). Then the samples or debris are collected and kept in an airtight container such as paint can or fruit jar to prevent evaporation of volatile liquids.

The forensic chemist uses different kind of methods in order to recover and identify the accelerants involved in the collected sample. Headspace technique is one of the most popular methods, in which a portion of the sample is taken in a clean and dry paint can or glass jar having a small hole in its lid. This hole of the lid is covered with a silicon septum glued on with super glue. The airtight container is then heated to vaporise the volatile residue present in the collected sample. A few millimetres of the vapour is taken by a syringe and tested it using gas chromatography.

Gas chromatography is a fast and simple analytical technique to characterize all type of organic materials including drugs, alcohol and poisons in blood or urine sample.

Experiment 8.3

Aim: Identification of arson accelerants by gas chromatography.

Principle: Arson accelerants can be hydrocarbons derived from petroleum such as gasoline or petrol, kerosene, and diesel. This experiment typically focuses on the identification of the accelerants involved in the collected sample or debris. For this purpose, standard samples of suspected accelerants such as gasoline, kerosene, and diesel are first analysed by gas chromatography and chromatograms are compared with that of the collected arson accelerants sample to identify the possible accelerant present in the collected sample.

Apparatus: Gas chromatograph, vials with caps, 5 µL syringe (for use with GC), 1 mL disposable syringe.

Chemical reagents: Gasoline, kerosene, and diesel.

Sample extraction and reference sample preparation: Headspace technique is used to extract the accelerant vapour from the fire debris sample. Take a portion of fire debris in a clean and dry conical flask and cover with a rubber septum. Then heat the conical flask on a water bath (80 °C) to vaporise the volatile residue absorbed in the collected sample. Using Hamilton gas-tight syringe withdraw 1 mL of the vapour for gas chromatography analysis.

For the preparation of reference samples, soak a wood chip in the accelerants for 4–6 h and then burnt it in a fume hood. After burning the total surface of the chip, extinguish the fire and keep it in a conical flask. Cover immediately with a rubber septum and withdraw 1 mL of the vapour using Hamilton gas-tight syringe for analysis.

Procedure: Take a 10 microliter (μL) syringe to withdraw the reference accelerant vapour sample. Pump the plunger for several times to fill the barrel of the micro-syringe. Inject the sample into injection port and record the chromatogram.

In a similar way, collect the vapour of the conical flask containing the debris and inject in the gas chromatograph injection port and record the chromatogram.

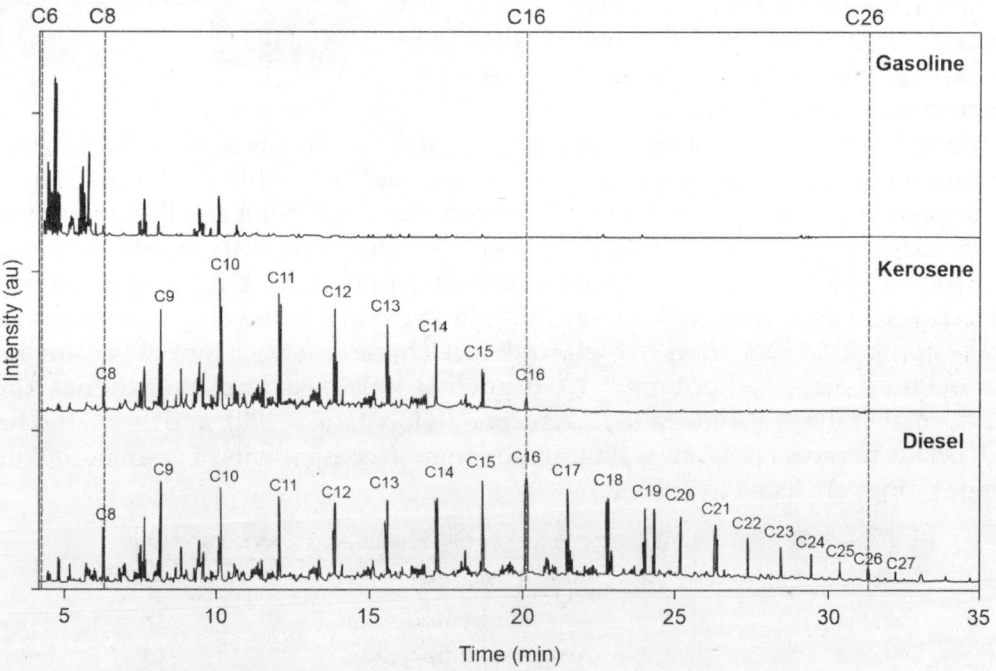

Fig. 8.5: GC chromatographs of refined gasoline, kerosene and diesel

Analysis of chromatograms and identification of accelerant: The accelerant present in the debris is identified by comparing the chromatogram of the sample with that of the standard samples (Fig. 8.5). The retention times and relative peak heights of the largest peak can provide the key information to identify the possible accelerant.

Every accelerant has a unique chromatogram characteristic. Gasoline is composed of hydrocarbons of low molecular weights containing 6–10 carbon atoms. Kerosene contains 8–16 carbon atoms containing hydrocarbons whereas diesel is comprised of wide variety of hydrocarbons having 8–28 numbers of carbon atoms. Therefore, the chromatograms of the accelerants are divided into three sections C_6–C_8, C_8–C_{16}, and C_{16}–C_{26} based on the molecular weight of hydrocarbons. The chromatogram of pure kerosene sample shows eight evenly distributed major peaks with few minor intermediate peaks whereas these eight peaks are also present in the chromatogram of diesel but in different proportions. In addition, diesel produces some evenly-spaced peaks at higher retention time originating from its higher boiling components. The petrol chromatogram shows multiple peaks of different proportions resulting from at least seven known components present in petrol in a certain relative proportion, discussed in next experiment.

8.3 ANALYSIS OF GASOLINE SAMPLE

Introduction: Gasoline or petrol is a colourless petroleum-derived flammable liquid that is used primarily as a fuel in most spark-ignited internal combustion engines. It consists mostly of organic compounds such as hydrocarbons obtained by the fractional distillation of petroleum. The largest fraction of gasoline is comprised of iso-octane while other aromatic compounds like toluene and benzene may also be included.

The nature of combustion of a fuel is largely influenced by its composition. A poorly formulated gasoline causes engine knocking and incomplete combustion. The combustion characteristics of a fuel are measured in a test motor and designated by an *octane number*. The octane rating of gasoline is a measure of its resistance to auto-ignition in spark-ignition internal combustion engines. Low octane numbers correspond to poor performance resulting in knocking of the engine whereas high octane numbers of fuel refers to high antiknock characteristics. Normal paraffin have low octane rating (*n*-heptane = 0), branched paraffins and naphthenes show intermediate ratings ('iso-octane', 2,2,4-trimethylpentane = 100), and aromatics have high octane numbers (toluene = 120). Some common components of gasoline and their octane ratings are listed in Table 8.1.

Table 8.1: Some common components of gasoline and their octane ratings		
Name	*Class of compound*	*Octane number*
Toluene	Aromatic hydrocarbon	120
Benzene	Aromatic hydrocarbon	106
2,2,4-trimethylpentane	Alkane (iso-octane)	100
Cyclohexane	Cyclic alkane	83
n-Pentane	Alkane	61.7
n-Hexane	Alkane	24.8
n-Heptane	Alkane	0
n-Octane	Alkane	19
n-Decane	Alkane	−41

Experiment 8.4

Aim: Determination of relative concentrations of different components in gasoline by gas chromatography.

Principle: Gasoline is a complex mixture of alkanes, cycloalkanes, aromatic hydrocarbons, and additives. This experiment typically focuses on the determination of relative concentrations of *n*-decane, *n*-octane, *n*-heptane, *n*-hexane, cyclohexane, toluene, iso-octane, and benzene. For this purpose, standards reference compounds are first analyzed by gas chromatography and the chromatograms are compared with that of gasoline samples. From the obtained relative percentage of these components in gasoline, the theoretical octane rating can also be calculated.

Apparatus: Gas chromatograph, vials with caps, 5 µL syringe (for use with GC), 1 mL disposable syringe.

Chemical reagents: n-decane, *n*-octane, *n*-heptane, *n*-hexane, *n*-pentane, toluene, iso-octane, benzene, cyclohexane and gasoline samples.

Sample preparation: Prepare the standard reference samples by adding 2 drops of each standard (*n*-decane, *n*-octane, *n*-heptane, *n*-hexane, toluene, iso-octane, benzene, and cyclohexane) with 2 mL of *n*-pentane in separate glass vials.

Procedure: For analysis of gasoline sample using a gas chromatography capillary column associated with a mass spectroscopic detector the following temperature profile can be used:

Equilibrium time: 0.50 min	Rate: 20 °C/min
Injector temperature: 150 °C	Final temperature: 90 °C
Initial oven temperature: 30 °C	Final time: 3.00 min
Initial time: 1.00 min	

Clean the injector syringe (5 µL) thrice before injecting a sample. Draw 3 µL of the liquid sample which is to be analysed and discard the liquid on paper towel by depressing the plunger.

Using the cleaned syringe, inject 1 µL of standard sample, i.e. known compound into injection port. Record the chromatogram as it is being plotted. The chromatogram is started with an air peak and ended with a flat, continuous baseline after each. After the standard samples, inject 1 µL of the gasoline sample into the injection port and record the chromatogram.

Calculation: Examine the chromatograms of the standard samples to identify the peak retention times and calculate the area covered under the peak. Identify and label the peaks in the chromatogram of the gasoline sample by comparing with known additives. However, additional peaks may be present in the chromatogram, as the gasoline sample contains so many additives.

Using the integration data (peak area) determine the % composition of the components (known additives) in the gasoline sample. Calculate the octane rating of the gasoline sample using the known octane ratings of the components and their relative percentage present in the gasoline sample.

BIBLIOGRAPHY

1. Cassidy Jr RF and Schuerch C. Gas chromatographic analysis of gasoline. A laboratory experiment. *Journal of Chemical Education*, 1976, 53(1), 51.

2. Kwon D, Ko MS, Yang JS, Kwon MJ, Lee SW and Lee S. Identification of refined petroleum products in contaminated soils using an identification index for GC chromatograms. *Environmental Science and Pollution Research*, 2015, 22(16), 12029–12034.

3. Sodeman DA and Lillard SJ. Who set the fire? Determination of arson accelerants by GC-MS in an instrumental methods course. *Journal of Chemical Education*, 2001, 78(9), 1228.

QUESTIONS

Multiple Choice Questions

1. The colour of phenolphthalein at pH range of 0–8.2 is:
   - (a) Pink
   - (b) Red
   - (c) Orange
   - (d) Colourless

2. At pH range of 8.2–10.0, phenolphthalein (H_2In) gives pink colour due to presence of:
   - (a) H_3In^+
   - (b) H_2In
   - (c) In^{2-}
   - (d) $In(OH)^{3-}$

3. At pH 9.0–12.0, phenolphthalein produces two absorption bands at ~280 and 555 nm due to ...... and ...... transition, respectively.
   - (a) π-π^*, n-π^*
   - (b) n-π^*, π-π^*
   - (c) π-π^*, σ-σ^*
   - (d) σ-σ^*, n-π^*

4. The arson accelerants are easily identified by:
   - (a) Paper chromatography
   - (b) Thin layer chromatography
   - (c) Ion-exchange chromatography
   - (d) Column chromatography

5. The octane numbers of *n*-heptane and iso-octane are considered to be:
   - (a) 0, 50
   - (b) 0, 100
   - (c) 100, 0
   - (d) 50, 0

Answers

1. (d); 2. (c); 3. (a); 4. (d); 5. (b)

Practice Questions

1. Define forensic sciences.
2. Discuss how phenolphthalein is used in trap cases.
3. Write down the principle of qualitative detection of phenolphthalein by TLC.
4. What is the working principle of spectrophotometric detection of phenolphthalein?
5. What is arson? Give some examples of arson accelerants.
6. Describe headspace technique of collecting the arson accelerants from a place of incident.
7. Briefly describe the working principle of identification of arson accelerants by gas chromatography.
8. What is octane number of gasoline? How it is related to the engine performance?
9. Briefly describe the working principle of the determination of relative concentrations of different components in gasoline by gas chromatography.

10. A GC chromatogram of gasoline containing hexane, pentane, toluene, octane and isopentane is depicted below. Identify the components A to E with explanation.

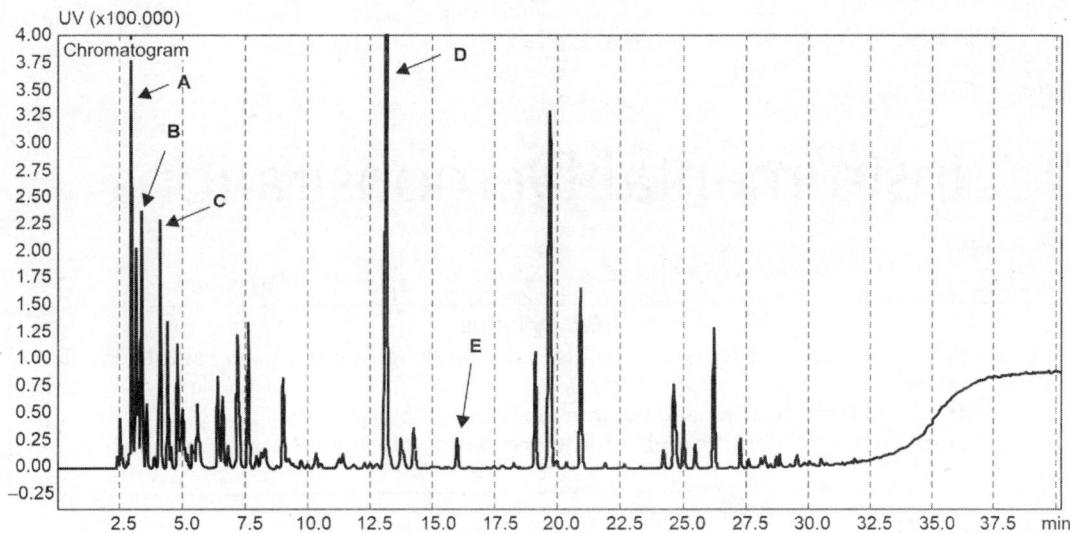

[Hint: Under isothermal conditions, the adjusted retention times for normal alkanes increase linearly with molecular weight (MW). Larger molecules (with higher MW and surface area) experience increased van der Waals (vdW) interactions which results in higher boiling point (BP) and thereby a longer retention time. For isomeric molecules the surface area, vdW force and BP decrease with increased branching. Therefore, a branched alkane has a lower retention time than its normal alkane with same MW.]

9

Instrumental Demonstrations

UGC Syllabus

1. Estimation of macronutrients: Potassium, calcium, magnesium in soil samples by flame photometry.
2. Spectrophotometric determination of iron in vitamin/dietary tablets.
3. Spectrophotometric identification and determination of caffeine and benzoic acid in soft drinks.

INTRODUCTION

The new CBCS syllabus includes flame photometry and UV-vis spectroscopy; this chapter only covers those instrumental methods with their principle, instrumentation and working procedure.

9.1 FLAME PHOTOMETRY

Flame photometry is widely employed tool in analytical chemistry. It is one kind of atomic emission spectroscopy (AES) and often known as flame emission spectroscopy. The IUPAC committee for spectroscopic nomenclature termed this spectroscopic method as flame atomic emission spectrometry (FAES). In 1980, David Richardson, Bowling Barnes, Robert Hood and John Berry first developed the technique in order to measure the Na^+ and K^+ ion concentrations. Later this technique is widely utilized to measure the concentration of alkali metals or alkaline earth metals, for example, Na, K, Li, Ca and Cs. The flame photometry involves atoms of the corresponding metal ions. Unlike atomic absorption spectroscopy (AAS), it measures the emission intensity of the metal ion introduced into the photometer. The wavelength of the colour is characteristic of the metal and the intensity of emission is a measure of the concentration of the element present in the sample. Thus, the technique is suitable for qualitative as well as quantitative analysis. Although the flame photometry and AAS techniques follow similar working principle, the two methods differ widely in various aspects (Table 9.1).

Principle: Flame photometry method is mostly used for elements which can be easily excited. Alkali and alkaline earth metal ions when inserted into flame transform into atoms. The flame serves dual purpose: It acts as an atomizer as well as an excitation source. The atoms get excited to higher energy levels and become unstable. These excited atoms while returning to the ground state emit radiations. The emitted radiations are generally in the visible region and each metal possesses a characteristic wavelength of emission (Table 9.2). The intensity of emitted radiation is directly proportional to the number of emitting atoms which in turn is proportional to concentration of the corresponding metal ion in the sample.

Table 9.1: Difference between flame photometry and atomic absorption spectrometry (AAS)

Flame photometry	Atomic absorption spectrometry
i. Atoms absorb the thermal energy and get excited, these excited atoms return to ground state with emission of radiation	i. The atoms which are unexcited in the ground state absorb the radiation
ii. This technique measures the intensity of emitted radiation	ii. This technique measures the intensity of absorbed radiation
iii. Intensity of emission depends on the number of atoms excited	iii. Intensity of absorption is dependent on the number of unexcited atoms present in the ground state
iv. It depends on the temperature of the flame	iv. It is independent of the temperature of the flame
v. Separate radiation source is not required	v. Separate radiation source is required
vi. Interference of spectral lines takes place	vi. There is no interference of spectral lines
vii. Lambert-Beer's law is not applicable	vii. Lambert-Beer's law is strictly applicable
viii. It is less expensive	viii. It is more expensive

Instrumentation: A flame photometer is a combination of several subunits (Fig. 9.1). The basic components of the instrument are discussed in brief:

(I) *Flame source:* A burner is used as source of the flame in flame photometer. It can maintain a constant temperature which is very critical factor in flame photometry. The flame acts as an atomizer as well as excitation source. It is produced by the chemical reaction between fuel and oxidant. Commonly, natural gas and air are utilized as fuel and oxidant, respectively, in flame photometry.

Table 9.2: Metal ions and their corresponding emitted colours

Element	Emitted wavelength	Flame colour
Sodium	589 nm	Yellow
Potassium	766 nm	Violet
Barium	554 nm	Lime green
Calcium	622 nm	Orange
Lithium	670 nm	Red

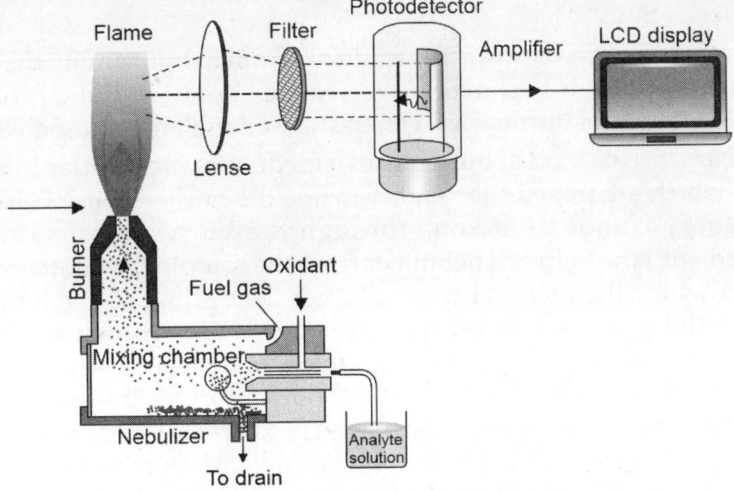

Fig. 9.1: Schematic diagram of flame photometer

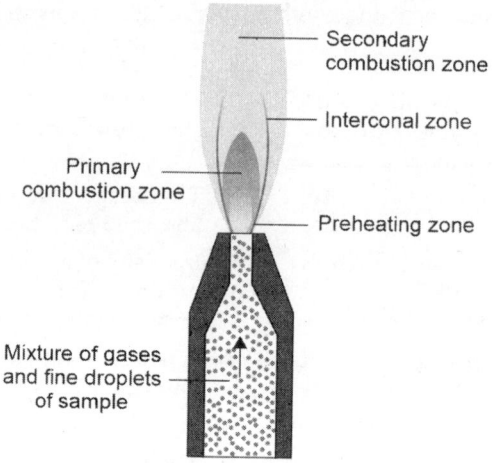

Fig. 9.2: Different regions of a flame

Temperature of the flame is controlled by the nature of the fuel and oxidant used and their relative proportions. The different types of fuel, oxidant and the temperature of the corresponding flames are listed in Table 9.3. The flame generally contains three basic regions, such as the primary combustion zone, interconal zone and secondary combustion zone (outer cone) as shown in Fig. 9.2. The ratio of fuel to oxidant determines the appearance and relative sizes of these regions.

Table 9.3: Flame components and corresponding flame temperatures

Fuel	Oxidant	Temperature, °C
Natural gas	Air	1700–1900
Natural gas	Oxygen	2700–2800
Hydrogen	Air	2000–2100
Hydrogen	Oxygen	2550–2700
Acetylene	Air	2100–2400
Acetylene	Oxygen	3050–3150
Acetylene	Nitrous oxide	2600–2800

(II) *Nebulizer:* The sample solutions are converted to aerosols by a jet of compressed gas using a nebulizer. The process is called nebulization where thermal vaporization and dissociation of aerosol particles at high temperature produce particles of small size and high residence time.

(III) *Atomizer burners:* The aerosol is introduced from nebulizer into the burner flame at a balanced rate where it is converted to atomic vapour and the process is called atomization. Two types of burners are generally used in flame photometry.

(a) *Premix burner:* This kind of burner uses a mixture of fuel, oxidant and the analyte sample which are mixed thoroughly inside the burner house before they enter the primary combustion zone through the burner ports. The solution is aspirated with the help of a nebulizer from the sample container into the mixing chamber where the fuel is also introduced. Depending on the number and size of the outlet port premix burners are classified as **Bunsen, Méker,** or **slot burners**. Bunsen burner has one large hole while Méker burner has a number of small holes. Slot burner has a slot as the outlet port for the gas mixture. Premix burners produce quiet flames which is easy to operate and the signals obtained in analysis are less noisy than that of total consumption burner.

(b) *Total consumption burner:* A total consumption burner is actually a combination of a nebuliser and a burner (Fig. 9.3). The mixture of fuel and oxidant being released at the outlet creates suction in the inner capillary which directly withdraws the sample and introduces it into the flame. The sample thus introduced contains suspended particles along with large droplets of solvents. Hence, the name of the burner is '**total consumption burner**'. As fuel and oxidants coming from separate ports get mixed above the burner orifices through turbulent motion, this kind of burners produce turbulent flames. IUPAC suggests replacement of the term 'total consumption burner' by the term '**direct-injection burner**'as the aspirated sample directly enters the flame.

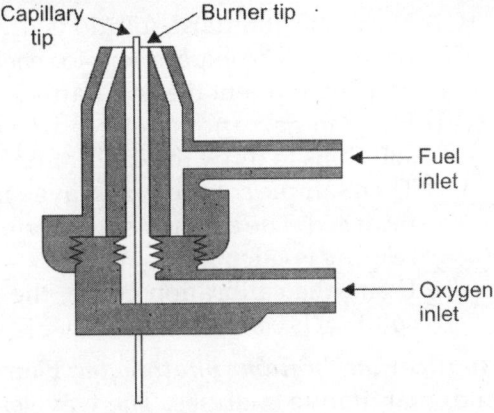

Fig. 9.3: Total consumption burner

(IV) *Optical system:* This part of flame photometer consists of convex mirror and lens. The mirror transmits the emitted radiation from the metals and helps in focusing the same on the convex lens. The lens helps the emitted light to focus on a slit or point.

(V) *Colour filters or monochromator:* The reflected light from the lens travels through the slit and reaches the filter where the relevant emissions are isolated. The filters are of two types: Absorption filter and interference filter. Absorption filters are limited in the visible range of spectrum whereas interference filters can be used in infrared (IR), visible (vis), and ultraviolet (UV) regions. These filters absorb certain fractions of the spectrum and transmit the wavelengths corresponding to the analyte element.

(VI) *Photodetector:* The detector is employed to evaluate the emission intensity. It converts the emitted radiation to electrical signals using a galvanometer. The electrical signal intensity is generated in direct proportion to the emitted light intensity.

(VII) *Amplifier and readout device:* The output signals coming from the detector is suitably processed in an amplifier and displayed on a readout device generally a galvanometer for flame photometry. Recent flame photometers are provided with microprocessor controlled electronics which produce outputs compatible with printers and computers, thus minimising chances of operator error during data transfer.

Procedure: The general procedure of measurement of concentration of different metal ions is described below.

(i) Sample and standard solutions are prepared with freshly prepared DI water.

(ii) The photometer is calibrated using varying ratio of air and gas injected to the flame. The flame is then allowed to be stable for about 5 to 10 min.

(iii) The photometer is switched on. Filter chamber is opened and appropriate colour filters are inserted.

(iv) Zero reading is adjusted on the galvanometer by spraying DI water into the flame.

(v) To adjust the sensitivity of galvanometer, most concentrated standard solution is introduced into the flame to get full deflection.

(vi) Again the galvanometer reading is adjusted to zero by spraying DI water into the flame.

(vii) Now, the standard solutions are introduced into the flame and the data are recorded. The experiment for each standard is performed in triplicate and after each experiment the apparatus is thoroughly washed with DI water.

(viii) Plot the galvanometer readings against the concentration of the standard solutions to draw the calibration curve.

(ix) Then sample solution is sprayed into the flame and the galvanometer reading is recorded. The experiment is performed in triplicate and the mean galvanometer reading is calculated.

(x) Using the calibration curve, the concentration of the elements in the sample solution is calculated.

Applications of flame photometer: Flame photometry is employed in both qualitative and quantitative analyses. The wavelength of the radiation emitted by the flame is characteristic of the elements used in the flame (qualitative property) and the intensity of emitted light is proportional to the amount of the element present in the injected sample (quantitative property). The technique is very useful in determination of alkali metals and alkaline earth metals which is very crucial for analysis of soil health and thereby choosing appropriate fertilizer for the soil.

The Na^+ and K^+ ions also play important role in numerous metabolic processes in human biology. The concentration of these ions in the body fluid can be determined by using a dilute blood serum sample in flame photometry.

The metal ion concentration in drinks and beverages can also be determined using flame photometry.

Advantages of Flame Photometer

- The flame photometry technique is quite easy and cost-effective.
- This technique is fast, suitable as well as possesses high selectivity and sensitivity.
- Both qualitative and quantitative analysis can be carried out by this method.
- Very low concentrations (ppm or ppb) of metal ions in sample can be detected by this method.
- The effect of interfering materials can easily be overruled by this method.

Disadvantages of Flame Photometer

Despite of the above mentioned advantages, the flame photometry method suffers from several disadvantages:

- The higher concentrations of metal ion in solution cannot be determined accurately.
- Only liquid samples can be performed. Sometimes, for the preparation of liquid samples lengthy steps are required.
- As natural gas and air flame is used as excitation source, the temperature of the flame is not high enough to excite transition metals. Therefore, the technique is not applicable for transition metals. Flame photometry is selective to alkali and alkaline earth metals.
- Certain disadvantages of this technique originate due to the low temperature. The flow rates, purity of the fuel and oxidant, aspiration rates and solution viscosity affect the interference and the stability of the flame and aspiration condition. So, it is very important to determine the emission of the unknown and standard solutions under the same conditions.

- Inert gases cannot be directly detected by this technique.
- Due to the non-radiating nature, carbon, hydrogen and halides are not detected by this technique.
- With this method, the molecular structure of the compound cannot be analysed.

9.2 ESTIMATION OF MACRONUTRIENTS: POTASSIUM, CALCIUM, IN SOIL SAMPLES BY FLAME PHOTOMETRY; MAGNESIUM IN SOIL SAMPLES BY ATOMIC ABSORPTION SPECTROSCOPY

Plant takes up the mineral nutrients from the soil. These nutrients can be categorized into two groups: Micronutrients and macronutrients. The micronutrients are the nutrients which are required in trace amounts for proper growth of the plants. It includes iron (Fe), manganese (Mn), copper (Cu), boron (B), zinc (Zn), molybdenum (Mo), and chlorine (Cl).

On the other hand macronutrients with most significant quantity are very essential for plant growth. Amongst them, nitrogen (N), phosphorus (P) and potassium (K) are required in very high concentration and these are called primary macronutrients. Due to high intake of these primary macronutrients by plants, soil shows deficiency of these ions. These deficiencies can be corrected by the addition of commercial or organic fertilizers. Secondary macronutrients such as calcium (Ca), magnesium (Mg), and sulfur (S) are also needed for sustained plant health, but in lower quantities than the primary macronutrients.

Plant absorbs the macronutrients in the form of ions from soils through the root. To identify the fertility level of the soils, the estimation of these ions is very essential. In this aspect, the more accurate and faster flame photometric method is preferred to the classical methods like gravimetric, colorimetric, or volumetric method for the determination of the concentration of cations in soils. In flame photometric method, the ions in the sample are atomized and led to the burner where atoms are exited. The emitted light during the relaxation of the atoms are detected by photodetector. Since the emitted light intensity of each element depends on the concentration of corresponding atoms in the flame which in turn depend on the concentration of the elements in solution. Measurement of the light intensity generated by a given element helps in quantitative determination of that element in the solution.

Experiment 9.1

Aim: Determination of calcium (Ca) in soil samples by flame photometry method.

Principle: The total amount of calcium is determined by measuring the percent transmittance (%T) of the soil sample using a flame photometer with help of a reference solution.

Apparatus: Conical flask, Whatman filter paper, pipette, measuring cylinder, volumetric flask, funnel, beakers.

Chemical reagents:

  (i) 0.1 N and conc. HCl

  (ii) Anhydrous calcium carbonate

(iii) *Standard reference solution of Ca^{2+}:* Prepare a standard calcium stock solution by dissolving 2.4973 g anhydrous calcium carbonate in mixture containing 500 mL DI water and 8 mL conc. HCl. Dilute the solution to 1 L with 0.1 N HCl in a volumetric flask to get the 1000 ppm Ca stock solution.

From this 1000 ppm Ca stock solution prepare another five standard solution containing 0 (blank), 50, 100, 150, and 200 ppm of Ca. For this purpose, transfer 0, 25, 50, 75 and 100 mL solutions of the 1000 ppm stock, respectively, to five different 500 mL volumetric flasks and dilute each solution to the mark of the volumetric flask with 0.1 N HCl.

Preparation of soil extract: 2.5 g air dried soil is taken with 50 mL 0.1 N HCl in a 125 mL conical flask. Using an automatic shaking machine, this mixture is agitated for 15 min. Then the soil suspension is filtered off using Whatman filter paper and the filtrate is collected in a 50 mL conical flask. This sample stock solution is directly used without further dilution for determination of Ca.

Procedure: The concentration of calcium in the sample stock solution is determined by means of a flame photometer. Measure the percent transmittance (%T) of each of the standard calcium solutions (50–200 ppm) and the blank solution. Draw a calibration curve by plotting the %T *vs* calcium concentration of the reference solutions. Then measure the percent transmittance (%T) of the soil extract. Finally, the amount of Ca in the soil solution extract is calculated by referring to the calibration curve.

Experiment 9.2

Aim: Determination of magnesium (Mg) in soil samples by atomic absorption spectrophotometer (AAS).

Principle: The total amount of magnesium is determined by measuring the percent absorbance of the soil sample using an atomic absorption spectrophotometer with the help of a reference solution.

Apparatus: Conical flask, Whatman filter paper, pipette, measuring cylinder, volumetric flask, funnel, beakers.

Chemical reagents:

(i) 0.1 N and conc. HCl

(ii) Mg metal

(iii) *Standard reference solution of Mg:* Prepare a standard magnesium stock solution by dissolving 1 g Mg metal in mixture containing 400 mL DI water and 20 mL conc. HCl. Dilute the solution to 1 L with 0.1 N HCl in a volumetric flask. Then transfer 100 mL of this 1000 ppm Mg solution to a 1 L volumetric flask and dilute the solution up to the mark of the volumetric flask with DI water to prepare the 100 ppm Mg stock solution.

From this 100 ppm Mg stock solution prepare another five standard solutions containing 0, 5, 10, 15, and 20 ppm of Mg. For this purpose, transfer 0, 25, 50, 75 and 100 mL solutions of the 100 ppm stock, respectively, to five different 500 mL volumetric flasks and dilute each solution up to the mark of the volumetric flask with 0.1 N HCl.

Preparation of soil extract: See the extraction procedure in Experiment 9.1.

Procedure: The concentration of magnesium in the sample stock solution is determined by atomic absorption spectrophotometer (AAS). Measure the percent absorption of each of the standard Mg solutions (5–20 ppm) and the blank solution. Draw a calibration curve by plotting the absorbance *vs* magnesium concentration of the reference solutions. Then measure the percent absorption of the soil extract. Finally, the amount of Mg in the soil solution extract is calculated by referring to the calibration curve.

Experiment 9.3

Aim: Determination of potassium (K) in soil samples by flame photometry method.

Principle: The total amount of potassium is determined by measuring the percent transmittance (%T) of the soil sample using a flame photometer with help of a reference solution.

Apparatus: Conical flask, Whatman filter paper, pipette, measuring cylinder, volumetric flask, funnel, beakers.

Chemical reagents:
   (i) 0.1 N and conc. HCl
  (ii) Anhydrous KCl
 (iii) *Standard reference solution of potassium (K):* Prepare a standard potassium stock solution by dissolving 0.953 g anhydrous KCl in 100 mL DI water. Dilute the solution to 1 L with 0.1 N HCl in a volumetric flask to get the 500 ppm potassium stock solution.

From this 500 ppm K stock solution prepare another five standard solutions containing 0, 5, 10, 20, and 25 ppm of K. For this purpose, transfer 0, 5, 10, 20, and 25 mL solutions of the 500 ppm stock, respectively, to five different 500 mL volumetric flasks and dilute each solution up to the mark of the volumetric flask with 0.1 N HCl.

Preparation of soil extract: See the extraction procedure in Experiment 9.1.

Procedure: The concentration of potassium in the sample stock solution is determined by flame spectrophotometer. Measure the percent transmittance (%T) of each of the standard potassium solutions (5–25 ppm) and the blank solution. Draw a calibration curve by plotting the %T *vs* potassium concentration of the reference solutions. Then measure the percent transmittance (%T) of the soil extract. Finally, the amount of potassium in the soil solution extract is calculated by referring to the calibration curve.

9.3 UV-VIS SPECTROPHOTOMETRY

The UV-vis spectrophotometry is one kind of absorption spectroscopy which deals with visible and adjacent UV range of electromagnetic radiations. The interaction of this region of electromagnetic radiation causes electronic transition within a molecule. Thus, the molecules interacting with the radiation absorbs radiation of wavelength corresponding to the electronic transition and transmittance is measured [A= log (T)]. The spectroscopy is based on Lambert-Beer's law.

$$A = \varepsilon cl$$

where, ε = the molar absorption coefficient or molar absorptivity
          l = the optical path length,
and c = the concentration of the analyte.

UV-vis spectroscopic technique is very fundamental and widely used analytical tool for characterization of numerous chemicals across the world. This technique is economical and easy to handle.

Principle: The working principle of UV-vis spectroscopic technique is the interaction of visible and nearby UV radiations with matters. When radiation of this region is passed through a dilute solution of a compound, the molecules absorb part of the radiation. This results in the electronic transition from ground state to excited state. Molecules with electrons available in π-bonding and/or non-bonding orbitals absorb energy which excites the electrons from these orbitals to higher antibonding orbitals. Four basic types of transitions are possible: π-π^*, n-π^*, σ-σ^* and n-σ^*. The more loosely bound electrons will be excited easily and thereby will absorb longer wavelength radiations. The energy of transition follows the order: σ-σ^* > n-σ^* > π-π^* > n-π^*. The light absorption by the compound produces a distinct spectrum which is characteristic of the compound.

Instrumentation: The spectrophotometer instruments operate with electromagnetic radiations in the visible and near UV range. Commercially available spectrophotometers are of two types: Single beam and double or split beam having different type of optical system. In general a UV-vis spectrophotometer comprises of the following parts (Fig. 9.4):

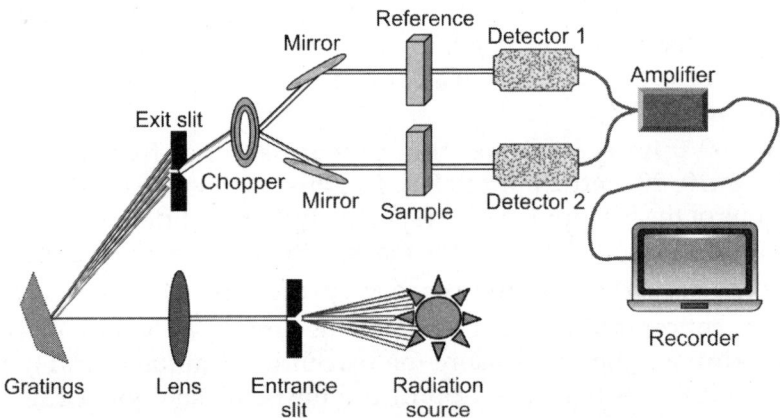

Fig. 9.4: Schematic representation of UV-vis spectrophotometer

(I) *Source of light:* The light source used in the UV-vis spectrophotometer should have some desirable properties such as low cost, a long service life, stability over time and most importantly brightness across a wide range of wavelength. There is no single light source that possesses all these qualities. The most commonly used light sources are tungsten lamp (for visible and near IR regions) and deuterium lamps (for UV region). Apart from these two light sources, sometimes xenon flash lamps are also used.

(II) *Monochromator:* Monochromators generally is composed of two segments: Dispersive element and slits.

The dispersive elements split the white light to its components of varying wavelengths. Spectrophotometers use prisms or diffraction gratings as dispersive elements. Previously prisms were very commonly used in spectrophotometers but nowadays diffraction gratings have received widespread usage. Generally used diffraction gratings consist of several hundred to ~2,000 equally spaced parallel grooves per millimeter length of the grating.

When white light is incident on the diffraction grating, because of interference, the light gets dispersed in a direction perpendicular to the grooves and light components of specific wavelengths are reflected only in specific directions (Fig. 9.5). These separated lights are then passed through the slit which selects the beam of monochromatic radiation having a single wavelength. In double beam spectrophotometers, the beam selected by the slit is further split into two beams.

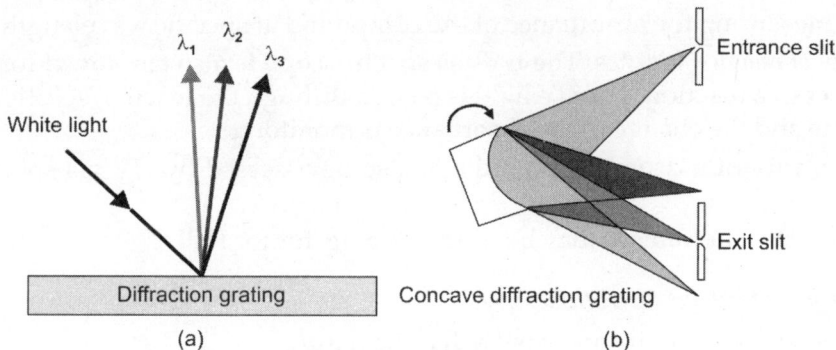

Fig. 9.5: Different type of diffraction gratings used in UV-vis spectrophotometer

(III) *Sample compartment:* The two beams from the slit enter the sample compartment where two cell holders are placed; one for sample and another for reference solution. The solutions are taken in square cells called cuvettes with optical path length of 1 cm. The cuvettes are made of either quartz or silica. Glass cuvettes are not suitable as it absorbs radiations in the UV region.

(IV) *Detector:* The next part of the spectrophotometer is detector. After passing through the sample compartment the light beams enter the detector. Generally, silicon photodiodes and photomultipliers are used as detectors for visible and ultraviolet regions.

In general double beam UV spectrophotometer uses two separate photocells to receive the beams from the sample cell and the reference cell. The intensity difference between the beams passing through sample and reference cell generates alternating currents in photocells.

(V) *Amplifier:* The amplifier receives the current signals from photocells. Then it uses a servometer to amplify the low intensity current signals to manifold in order to get recordable signals.

(VI) *Recording devices:* A pen recorder connected to the amplifier records the data and a computer associated with the pen recorder stores all the data, processes it and produces the absorption spectrum of the compound.

Applications of UV-vis Spectroscopy

• *Qualitative and quantitative analysis:* UV absorption spectroscopy is mainly used for characterization of organic compounds which absorbs UV radiation. The qualitative analysis is performed by recording the absorption spectra of the compound and comparing the same with the spectrum of known compound. This spectroscopic technique is also helpful in quantitative determination of compounds.

• *Structure elucidation:* This spectroscopic technique is useful in determination of the structure of organic molecules. The presence of unsaturation or heteroatoms can be assessed by studying the absorbance spectra of the compound.

• *Impurity detection:* The UV-vis spectrophotometric method is widely used in the purity checking of various organic molecules. Every compound has its characteristic absorption spectra and the impurities can be identified from the additional peaks observed in the UV-vis spectrum. The alternative method of impurity detection involves measuring the absorbance of the compound at specific wavelength.

• *Study of reaction kinetics:* The UV-vis spectroscopy is also employed for studying the kinetics of a reaction. The UV light is passed through the reaction solution taken in the cuvette and the change in the absorbance is monitored.

• The purity of a drug compound can also be assessed by UV-vis spectrophotometry.

• UV spectrophotometer may be used as a detector for HPLC.

Advantages of UV-vis Spectroscopy

• Its fast analysis ability and easy to handle nature.
• It is fairly inexpensive instrument.
• It shows high sensitivity and low signal to noise ratio and simple sample compartments.
• Require very small volume of sample for analysis.
• Linearity over wide range of concentration for dilute solutions.

Disadvantages of UV-vis Spectroscopy

• The broad nature of the absorption spectra sometimes makes the analysis of unknown sample difficult.
• It cannot provide information about the functional groups. The absorbance peak is observed when the functional group is attached to conjugation.
• For concentrated solutions results are not always linear.
• The molecules which do not absorb light at this wavelength region cannot be measured with this technique.
• High voltage is required for initiation.

9.4 SPECTROPHOTOMETRIC DETERMINATION OF IRON IN VITAMIN/DIETARY TABLETS

Experiment 9.4

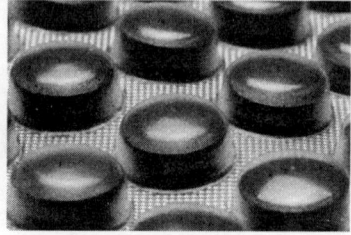

Aim: Quantitative determination of iron content of a commercially available vitamin tablet using UV-vis spectrophotometry.

Principle: In this experiment, an iron containing vitamin tablet is dissolved in a strong acid, and the cooled mixture is filtered into a volumetric flask. After that the Fe^{3+} in the mixture is quantitatively reduced to Fe^{2+} using hydroquinone. This freshly-dissolved colourless Fe^{2+} ion forms intense red colour complex with 1,10-phenanthroline or *o*-phenanthroline (Fig. 9.6).

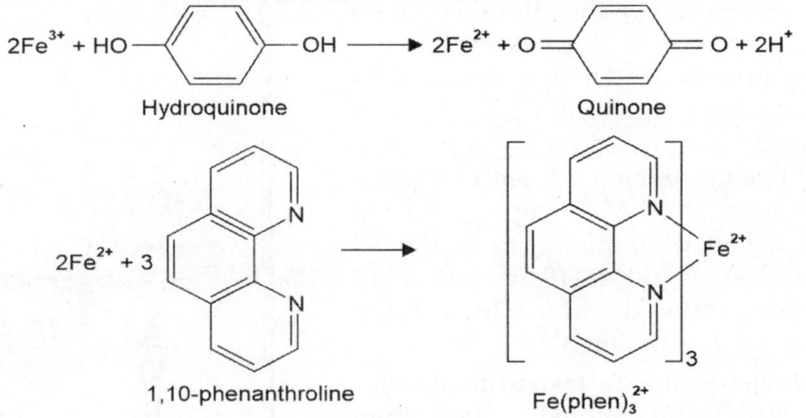

Fig. 9.6 Reduction of Fe^{3+} to Fe^{2+} and formation of intense red colour complex, $Fe(phen)_3^{2+}$

At pH = 3 or higher value, this complex is stable. The red colour complex, $Fe(phen)_3^{2+}$, has a maximum absorption at ~510 nm. Absorbance measurement of the analyte solutions at λ_{max} (~510 nm) is the key step for determination of iron concentration in the sample.

A series of standard solutions of $Fe(phen)_3^{2+}$ of known concentrations is prepared. Absorbance intensities of those solutions are measured and the values are plotted against concentration to obtain the calibration curve. By determining the absorbance of the sample solution, iron content of the vitamin sample can be calculated from the standard curve.

Apparatus: Pipette, measuring cylinder, volumetric flask, funnel, beakers, amber bottle, cuvettes, pH paper.

Reagents:

(i) *Hydroquinone:* Prepare an aqueous solution of hydroquinone with concentration of 2 g/L and store it in amber bottle.

(ii) *Sodium citrate:* Prepare an aqueous solution of sodium citrate with concentration of 50 g/L.

(iii) *1,10-phenanthroline:* Prepare 1 g/L solution in water, and store in amber bottles.

(iv) *Standard stock solutions (0.04 mg Fe per mL):* Dissolve 0.5602 g of reagent-grade $Fe(NH_4)_2(SO_4)_2 \cdot 6H_2O$ in a 2 L volumetric flask containing 2 mL of 98% H_2SO_4 and dilute to the mark with DI water.

Prepare four standard solutions with 10, 5, 2, and 1 mL aliquots of prepared Fe^{2+} stock solution and a blank solution without Fe^{2+} (i.e. using DI water only). All five solutions should be made directly in 100 mL volumetric flasks. In each case, pipette out the required volume of Fe^{2+} stock solution directly into a 100 mL volumetric flask. Then add sodium citrate solution to make the pH ~3.5. Add 10 mL of hydroquinone and 10 mL of 1,10-phenanthroline to each of the five volumetric flasks. Dilute all the solution up to the mark of the volumetric flask using DI water.

Preparation of sample: Collect a vitamin tablet containing iron, check the label and note down the iron content (mg of Fe per tablet). Take the tablet into a 100 mL beaker and add 25 mL of 6 M HCl into it. Gently heat the solution on a Bunsen burner flame for 15 minutes for complete dissolution (Fig. 9.7). Perform the experiment inside a fume hood. Then cool down the solution to room temperature and filter it directly into a 100 mL volumetric flask. To ensure complete quantitative transfer, washing of the beaker and filtration should be repeated for several times. Make the volume up to the mark with DI water and label this as 'solution A'. Using a graduated pipette take out 5 mL of solution A to another 50 mL volumetric flask and dilute (10 times dilution) it up to the mark with DI water. This diluted solution is marked as 'solution B'.

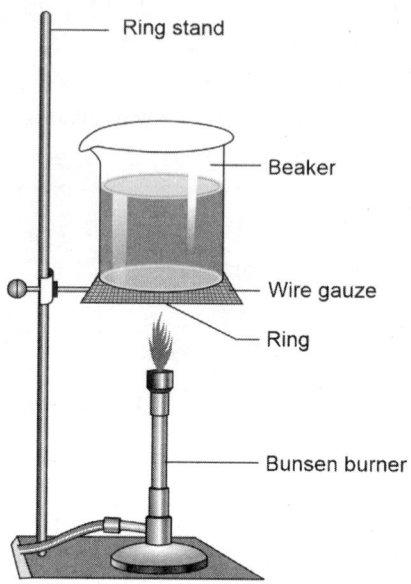

Using a volumetric pipette, take 5 mL of 'solution B' into a clean 100 mL volumetric flask. Then add sodium citrate solution dropwise to make the pH of the solution ~3.5. Check the pH with a pH paper or pH meter. After that, add 10 mL of hydroquinone solution followed by 10 mL of 1,10-phenanthroline solution. Dilute this mixture solution up to the mark (100 mL) with DI water (20 times dilution). Mix it well and label this mixture as 'unknown solution C'. The complex should form within five minutes.

Fig. 9.7: Dissolution of iron tablet

Procedure: The concentration of iron in the vitamin sample extract is determined by UV-vis spectrophotometer. Measure the absorbance of each of the standard Fe(phen)$_3^{2+}$ complex solutions (4×10^{-4} to 4×10^{-3} mg Fe per mL) and the blank solution. Draw a calibration curve by plotting the absorbance versus iron concentration of the reference solutions. Then measure the absorbance of the vitamin sample extract (unknown solution C). Finally, calculate the amount of iron in the vitamin sample extract by referring the absorbance value to the calibration curve.

9.5 SPECTROPHOTOMETRIC IDENTIFICATION AND DETERMINATION OF CAFFEINE AND BENZOIC ACID IN SOFT DRINKS

Experiment 9.5

Aim: Measurement of two major species such as caffeine and sodium benzoate in soft drinks using spectrophotometry.

Principle: The commercial soft drinks contain several ingredients such as carbonated water, high fructose corn syrup and/or sugar, caramel colour, phosphoric acid, artificial and natural flavours, sodium benzoate, caffeine,

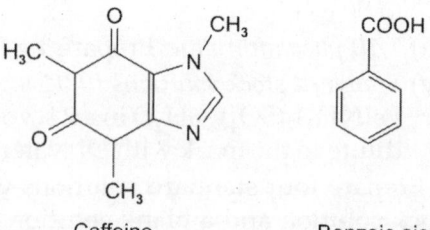

Fig. 9.8: Chemical structures of caffeine and benzoic acid

etc. Caffeine is a central nervous system stimulant of the methylxanthine class (Fig. 9.8). Benzoic acid (Fig. 9.8) is a food preservative which is widely used in acidic (pH 2.5–4) foods and it is generally added as salt, sodium benzoate. Even sometimes the benzoic acid naturally occurs in several fruits with very high amounts.

To avoid any kind of interference by some other UV active molecules like aspartame (a sugar substitute in diet soda) or caramelized sugar (colourant in darkly coloured drinks), white coloured non-diet soft drinks like Mountain Dew, Sprite, etc. are chosen for this experiment. Mountain Dew has high caffeine content whereas Sprite is a caffeine free soft drink. Although these colourless soft drinks have some UV absorbance from the matrix, the systematic error due to this is very small in comparison to the concentration levels of caffeine and benzoic acid.

Pure benzoic acid and caffeine produce two absorbance maxima at ~229 nm and ~272 nm, respectively. Soft drinks containing benzoic acid and caffeine will also record these absorbance maxima. By comparing the intensities with reference solutions the content of benzoic acid and caffeine in sample soft drinks are calculated.

Apparatus: Beaker, volumetric flask, measuring cylinder, filter paper.

Chemical reagents:

(i) *0.10 M HCl:* Dilute 8.2 mL of 37 wt % HCl to 1 L with DI water.

(ii) *Standard benzoic acid solutions:* Prepare 100 mg/L aqueous stock solution of benzoic acid and then make a set of standard solutions containing 2, 4, 6, 8 and 10 mg/L benzoic acid in 0.01 M HCl. To prepare these standard solutions, take 2, 4, 6, 8 and 10 mL of benzoic acid stock solution, respectively, in 100 mL volumetric flasks and add 10 mL 0.1 M HCl to each volumetric flask. Dilute each solution up to the mark with DI water.

(iii) *Standard caffeine solutions:* Prepare 200 mg/L aqueous stock solution of caffeine and then make a set of standard solutions containing 4, 8, 12, 16 and 20 mg/L caffeine in 0.01 M HCl. To prepare these standard solutions, take 2, 4, 6, 8 and 10 mL of benzoic acid stock solution respectively in 100 mL volumetric flasks and add 10 mL 0.1 M HCl to each volumetric flask. Dilute each solution up to the mark with DI water.

Sample preparation: Using a measuring cylinder take 20 mL of soft drink in a 100 mL beaker. Warm up the solution on a hot plate to expel CO_2 and when effervescence stops, filter the warm liquid. Cool down the solution to room temperature and pipette out 4 mL aliquot into a 100 mL volumetric flask. Add 10 mL of 0.10 M HCl and dilute up to the mark with DI water.

Procedure: Record the absorbance of blank (0.01 M HCl) from 350 to 210 nm and set this as baseline. Then collect the absorbance data for the two sets of reference solutions (benzoic acid and caffeine). Collect the readings in triplicate for each solution. Plot the calibration curves (absorbance *vs* molar concentration) for benzoic acid and caffeine by using the absorbance values of the corresponding standards solution at 272 and 229 nm, respectively. The molar absorptivity (ε) values of benzoic acid and caffeine can be calculated from the slope of the curves.

Record the absorption spectrum (Fig. 9.9) of the sample soft drink solution. From the absorbance values of the sample solution at 272 nm and 229 nm, find the concentrations of benzoic acid and caffeine, respectively, in the original soft drink. Run at least three trials for each of the prepared soft drink samples. Record the data in triplicate.

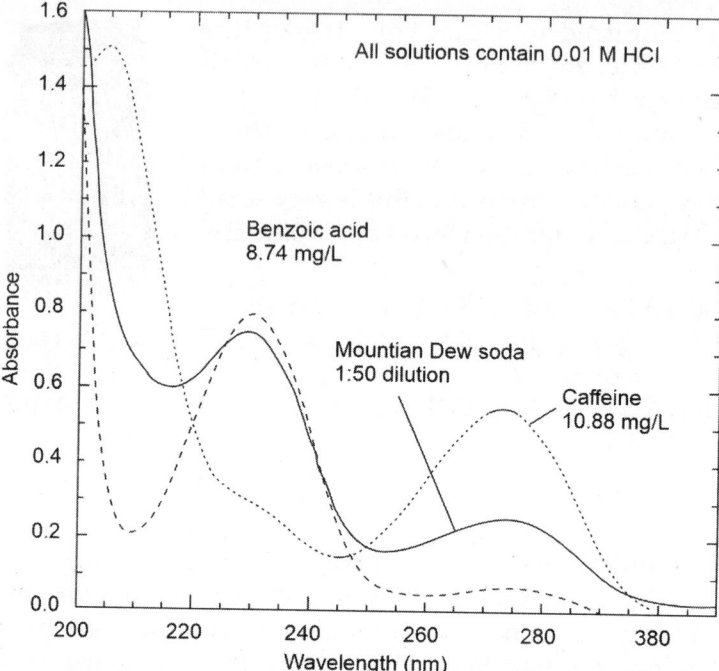

Fig. 9.9: Ultraviolet absorption of benzoic acid, caffeine, and a 1:50 dilution of Mountain Dew soft drink. All solutions contain 0.01 M HCl

Calculation: With the absorbance at 272 nm and 229 nm, determine the concentrations of benzoic acid and caffeine in sample solution using Lambert-Beer's equation.

BIBLIOGRAPHY

1. Atkins RC. Colorimetric determination of iron in vitamin supplement tablets. A general chemistry experiment. *Journal of chemical education*, 1975, 52(8), 550.
2. Brupbacher RH, Bonner WN, Sedberry Jr SE. Analytical methods and procedures used in the soil testing laboratory, 1968.
3. McDevitt VL, Rodriquez A and Williams KR. *J. Chem. Ed.* 1998, 75, 625.

QUESTIONS

Multiple Choice Questions

1. Flame photometry is a kind of:
   (a) Absorption spectroscopy　　　　　　(b) Emission spectroscopy
   (c) Fluorescence spectroscopy　　　　　(d) None of these
2. Which of the following statements is true about flame emission photometry?
   (a) Atoms in vapour state absorb radiation and get excited to higher energy states
   (b) Medium absorbs radiation and transmitted radiation is measured
   (c) Colour and wavelength of the flame is measured
   (d) Only wavelength of the flame is measured
3. In flame photometry, ... and ...... are measured for quantitative and qualitative analysis, respectively.
   (a) Intensity and colour　　　　　　　(b) Colour and intensity
   (c) Intensity and velocity　　　　　　(d) Colour and Frequency

4. In which of the following area flame photometry is not applicable?
   (a) Analysis of biological fluids
   (b) Determination of sodium, potassium in soil
   (c) Determination of metals such as Mn, Cu
   (d) Analysis of complex mixtures
5. Which is not a characteristic of laminar flow burner?
   (a) Noiseless
   (b) Stable flame for analysis
   (c) Efficient atomization of sample
   (d) Sample containing two or more solvents can be burned efficiently
6. Laminar flow burner used in flame photometry is also known as:
   (a) Turbulent burner   (b) Premix burner
   (c) Total consumption burner   (d) Nozzle mix burner
7. The advantage of using prism monochromators is:
   (a) Dispersion is non-overlapping
   (b) Dispersion occurs in non-linear manner
   (c) Dispersion is overlapping
   (d) Dispersion occurs in a linear manner
8. Which of the following is a characteristic of grating monochromators?
   (a) Dispersion is non-overlapping
   (b) Dispersion occurs in non-linear manner
   (c) Dispersion is overlapping
   (d) Dispersion occurs in a linear manner
9. Among the following which is not used as a detector in flame photometry?
   (a) Photonic cell   (b) Photovoltaic cell
   (c) Photoemissive tube   (d) Chromatogram
10. Given below is the diagram of flame emission photometers. Identify the unmarked (?) component.

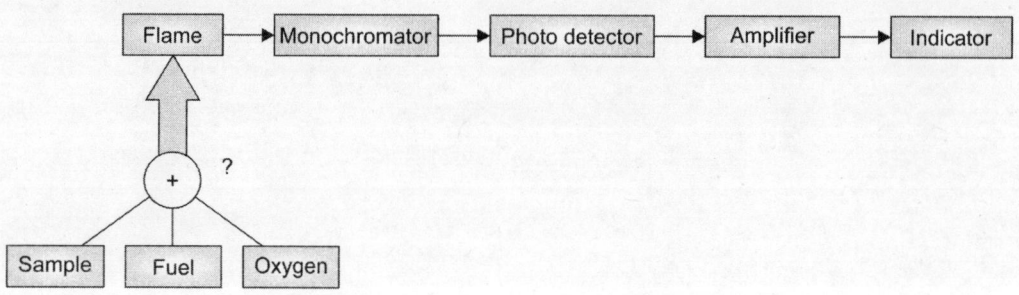

   (a) Filter   (b) Atomiser
   (c) Pressure regulator   (d) Burner
11. Energy of electronic transition follows the order:
   (a) $\sigma\text{-}\sigma^* > n\text{-}\sigma^* > \pi\text{-}\pi^* > n\text{-}\pi^*$
   (b) $\sigma\text{-}\sigma^* < n\text{-}\sigma^* > \pi\text{-}\pi^* > n\text{-}\pi^*$
   (c) $\sigma\text{-}\sigma^* < n\text{-}\sigma^* < \pi\text{-}\pi^* > n\text{-}\pi^*$
   (d) $\sigma\text{-}\sigma^* < n\text{-}\sigma^* < \sigma\text{-}\sigma^* < n\text{-}\sigma^*$

12. During spectrophotometric estimation of iron, which of the following is used to reduce Fe^{3+}?
    (a) 1,10-phenanthroline   (b) Hydroquinone
    (c) Water   (d) UV light
13. Benzoic acid and caffeine in soft drinks have absorbance maxima at ...... and ......, respectively.
    (a) 329 nm, 472 nm   (b) 472 nm, 329 nm
    (c) 229 nm, 272 nm   (d) 272 nm, 229 nm

Answers

1. (b); 2. (c); 3. (a); 4. (d); 5. (d); 6. (b); 7. (a); 8. (d); 9. (d); 10. (d); 11. (a); 12. (b); 13. (c)

Practice Questions

1. Discuss briefly the principle underlying the quantitative analysis by flame photometry.
2. What are the advantages and disadvantages of flame photometry?
3. Write a short note on atomizer burners used in flame photometry.
4. What are the applications of flame photometry?
5. What are the advantages and disadvantages of flame photometer?
6. What are the differences between flame photometry and atomic absorption spectrometry (AAS)?
7. Discuss briefly about micronutrients and macronutrients for plant growth with examples.
8. How to extract soil sample for analysis?
9. Discuss briefly the principle of UV-vis spectroscopy.
10. What are the advantages and disadvantages of UV-vis spectroscopy?
11. What are the different applications of UV-vis spectroscopy?
12. How do you spectrophotometrically identify and determine caffeine and benzoic acid in soft drinks?

Appendices

Some physical quantities and their units and symbols

Physical quantity	Units	Symbol
Concentration	Molarity	M
	Normality	N
Mass	Gram	g
	Milligram	mg
Volume	Litre	L
	Millilitre	mL
Temperature	Celsius	°C
	Kelvin	K
Time	Hour	h
	Minute	min
	Second	s
Length	Centimetre	cm

APPENDIX II
Some chemicals with their hazard labels according to globally harmonized system (GHS)

Chemicals	Formula	GHS hazard labels	CAS No.
Acetonitrile (ACN)	H_3CCN	Flammable, irritant	75-05-8
Aluminium trichloride	$AlCl_3$	Corrosive	7446-70-0
Ammonium hydroxide	NH_4OH	Corrosive, irritant, environmental hazard	1336-21-6
Boric acid	H_3BO_3	Health hazard	10043-35-3
Calcium oxide	CaO	Corrosive, irritant	1305-78-8
Carbon disulphide	CS_2	Flammable, health hazard, irritant	75-15-0
Carbon tetrachloride	CCl_4	Toxic substance, health hazard	56-23-5
Chloroform	$CHCl_3$	Toxic substance, health hazard	67-66-3
Ferric chloride	$FeCl_3$	Corrosive, irritant	7705-08-0
Hydrochloric acid	HCl	Corrosive, irritant, health hazard	7647-01-0
Nitric acid	HNO_3	Oxidizer, toxic substance, corrosive	7697-37-2
Potassium cyanide	KCN	Toxic substance, health hazard, corrosive, environmental hazard	006-007-00-5
Potassium iodide	KI	Health hazard	7681-11-0
Silver nitrate	$AgNO_3$	Oxidizer, corrosive, environmental hazard	7761-88-8
Sodium hydroxide	$NaOH$	Corrosive	1310-73-2
Sulfuric acid	H_2SO_4	Corrosive	7664-93-9
Zinc oxide	ZnO	Environmental hazard	1314-13-2
Benzoic acid	$C_6H_5CO_2H$	Corrosive, health hazard	65-85-0
Calcium chloride	$CaCl_2$	Irritant	10043-52-4
Potassium biiodate	$KHIO_3$	Oxidizer, corrosive	13455-24-8
Sodium carbonate	Na_2CO_3	Irritant	497-19-8
Toluene	$C_6H_5CH_3$	Flammable, irritant, health hazard	108-88-3
Caffeine	$C_8H_{10}N_4O_2$	Irritant	58-08-2
Hydroquinone	$C_6H_4\text{-}1,4\text{-}(OH)_2$	Health hazard, corrosive, irritant, environmental hazard	123-31-9

Index